Sydney Ferris Walker

Electricity in our Homes and Workshops

A practical treatise on auxiliary electrical apparatus

Sydney Ferris Walker

Electricity in our Homes and Workshops
A practical treatise on auxiliary electrical apparatus

ISBN/EAN: 9783337405779

Printed in Europe, USA, Canada, Australia, Japan

Cover: Foto ©Andreas Hilbeck / pixelio.de

More available books at **www.hansebooks.com**

ELECTRICITY

IN OUR

HOMES AND WORKSHOPS.

A PRACTICAL TREATISE ON
AUXILIARY ELECTRICAL APPARATUS.

BY

SYDNEY F. WALKER,

M.I.E.E., M.I.M.E., Assoc. M. Inst. C.E. M. Amer. I.E.E.

𝔚𝔦𝔱𝔥 𝔑𝔲𝔪𝔢𝔯𝔬𝔲𝔰 𝔍𝔩𝔩𝔲𝔰𝔱𝔯𝔞𝔱𝔦𝔬𝔫𝔰.

LONDON:

WHITTAKER & CO., PATERNOSTER SQUARE.

GEORGE BELL & SONS, YORK STREET, COVENT GARDEN.

1889.

PREFACE.

THE following pages have been written to supply a want which, it is believed, has been much felt. They are an attempt to explain in simple terms the ordinary every-day working of some of the forms of electrical apparatus that are in use by outsiders, and not under the constant supervision of trained electrical engineers; and also, it is hoped that they will be of service to young electrical engineers who are just commencing to make their practical experience.

It has been felt that a connecting link is wanting between the Electricity of the schools, and the electrical engineering of practical life.

In too many instances, practical men have been tempted to throw aside the study of the principles of the science altogether, because the book they consulted or the lecture they attended has failed to assist them in the elucidation of those practical problems which it was the business of their life to solve, and even in some cases would appear to have indicated a course of action or a method of treatment quite at variance with that which eventually proved successful.

As it has been the Author's lot to work, in the practical sense, upon all the apparatus he deals with; and as he has always found a careful study of the principles of the science of very great service to him, and further, as he has never found those principles misleading in any case, provided that all the conditions were accurately known, he ventures to hope that what he places before the reader, in the following pages, may be of some service to him.

The substance of the chapters which follow is, in fact, merely a reproduction of the Author's every-day intercourse with the pupils of his firm, and with those who have employed him in business.

From what has been said it will easily be understood, that this book makes no claim to supersede, nor to compete with, the able works of Ayrton, Thomson, Kapp, and others. If it has any relation to them at all, it is supplementary; but in fact, it is intended to hold a distinct place apart from those works, and to cater, so far as may be possible, for those to whom the meat given there would be too strong.

When the above was written, the Author hoped to have covered the whole ground occupied by electrical apparatus. Time, however, has obliged him to deal only with what are known as auxiliaries to the practical business of life, those in which only very small currents are used. Should the present work, however, find favour with those for whom it has been written, he hopes to deal with electric lighting, transmission of power, electrical measurements, and kindred subjects in another book. He has now only to tender his best thanks to those firms who have kindly lent illustrations for use in the book, viz.—to Messrs. Gent and Co., Leicester, Paterson and Cooper, Johnson and Phillips, Cox Walker and Co., New Telephone Co., Tasker Sons and Co., L. J. Crossley; and also to Messrs. A. S. Barnard and G. H. Lovegrove, who made the original drawings, and to Major Barnard, who has kindly assisted in correcting the proofs.

SYDNEY F. WALKER.

CONTENTS.

IX

LIST OF ILLUSTRATIONS.

xi

ELECTRICITY
IN OUR HOMES AND WORKSHOPS.

CHAPTER I.

INTRODUCTION.

It may be as well to define the terms that will be used in the course of this book, and to give explanations that will enable readers to follow the instructions without referring to another. At the same time, only what may be termed popular definitions will be given, the reader being referred to existing text-books, should he desire to follow the matter up further.

GLOSSARY OF TERMS.

What is Electricity? We do not know, and for practical purposes it is not necessary that we should know. We are only concerned in what its properties are— how we can make it obedient to our will. Certain points, however, may be noted. For instance, it is necessary to remember that we do not find electricity ready made for us, as we do coal. In order to generate electricity, as it is termed, work must be done by some body, or energy expended on that body; and this is true even of natural sources of electricity, such as those which give rise to the phenomena of thunderstorms; the only difference being that, in these cases, nature does the work, just as she has done the work of storing coal or placing the glacier.

So far as our knowledge goes at present, electrical energy, in the form of electric charge or electric current, is analogous to heat energy, light energy, chemical energy, etc.; and when electrical energy is generated, it has been produced by a transference, directly or indirectly, from one of the other forms of energy. Also, when electrical energy is apparently lost or dissipated, it is merely transferred into some other form of energy, in the same way that it was or could have been generated. The whole art of electrical engineering consists in the study of how best to transfer energy from its existing form to that of electrical energy, with as little escape to other forms as possible; and how to utilize the re-transference of electrical energy into some other form, as best to further the work in hand.

Electro-motive Force is analogous to pressure in steam or water; in fact, the term "electrical pressure" is coming into daily use. It is the property of doing electrical work, or of expending electrical energy, that may exist between any two points, such as the terminals of a dynamo, or of a galvanic battery; between a highly charged cloud and the earth, or another cloud. Whenever energy is expended upon a body in such a manner that electricity is generated, electrical engineers express the fact by saying that a certain electro-motive force is created; that is, a certain power of doing electrical work. Electro-motive force is sometimes expressed as *Difference of Electrical Potential*, in a manner analogous to difference of temperature; and this expression merely signifies, as before, that a measurable electrical force exists between two points. It is also sometimes referred to as *Tension*, or *Difference of Tension*, meaning the same thing.

The Volt is the unit of electro-motive force, electric tension, and electric potential. It is used by electrical engineers in the same way that the lb. is used by mechanical engineers. Electrical engineers talk of so many *Volts* between the terminals of a dynamo, or between electric lighting mains, just as mechanical engineers talk of so many pounds pressure in the steam boiler or the cylinder of the steam engine; and the two expressions are used in exactly the same manner. A mechanical engineer having to do certain work, will first ascertain the pressure of steam he can obtain before calculating the size of the engine he will require. So, an electrical engineer, having a certain amount of light to furnish, or certain mechanical work to do, will first ascertain what pressure or voltage he can depend on, before he calculates the sizes of his carbons or his motor.

The two expressions agree also in the fact that both become less, as they recede from the source, in proportion to the work they do. Every one is familiar with the fact that the steam issuing from the exhaust of a steam engine is at a lower pressure than at the entry port, and mainly because it has done work in passing through the engine.

So, where a certain voltage exists at a certain point, say between two electric light mains; when work is done, the voltage is always lower beyond the work, whether that be a lamp, a motor, or any other consumer of electric current. Electro-motive Force is usually written E.M.F.

The Volt is a perfectly measurable quantity, both in its effects and in its inception. It is taken from the attraction or repulsion between two electrically charged bodies, and consists of a definite multiple of the force exerted by one charged body of a certain size—the gramme—upon another of the same size at a certain distance—the centimètre. For practical purposes it will be sufficient for the reader to know that the Volt is very nearly equal to the electro-motive force furnished by the Daniell cell, and that the Le Clanché battery, the Sulphur-Sal-Ammoniac battery, and the Bichromate battery each give about 1½ Volt per cell, between their terminals, when not furnishing any current; that is, when no current is passing through them.

Electrical Resistance, Electrical Conductivity or *Conducting Power.*

All known substances allow of the passage of electricity through them ; but some with much greater ease than others. The metals, for instance, are said to be *Conductors*, as they offer comparatively small *Resistance* to the passage of electrical currents, in comparison with such substances as silk, cotton, glass,

earthenware, which have been termed Insulators, from
the supposition, in the early days of the science, that
they did not conduct at all. In reality there is an
irregular gradation from the bodies called insulators,
at one end of the scale, to those called conductors at
the other. Some substances,—such as carbon, sulphur,
the acids, water,—hold intermediate positions between
the so-called insulators and the so-called conductors.
Under certain conditions they may be classed as imperfect conductors, and under other conditions as imperfect
insulators. A few of them,—as carbon and others,—
have been termed semi-conductors.

The conducting power of, or the resistance offered by,
any group of bodies differs very considerably among
themselves. As, for instance, of the metals, silver and
copper conduct best :—

Representing the resistance of silver and copper by 1
That of Iron would be represented   „ 6
    „   Lead   „   „   „ 12
    „   German Silver   „   „ 12
etc., etc.

Using the same standard, carbon would be represented by from 1,500 to 40,000, dilute sulphuric acid
100,000, water 1,000,000; *dry* air by infinity.

The conducting power of all bodies is naturally
affected by their form and size ; thus, it can easily be
understood that it will require a greater expenditure
of energy to drive a given quantity of electricity
through two miles of wire of a given size than through
one mile. Further, the conducting power increases

with the size, or cross section.   Thus a No. 8 standard-wire gauge wire of a given length offers less resistance, or requires a smaller force to drive a certain quantity of electricity through it, than a No. 11 wire of the same length ; and a copper cable half an inch in diameter will have four times the resistance of a cable composed of similar metal and of the same length, one inch in diameter.

Thus it comes about that bad conductors may offer a low resistance, or demand the expenditure of a small electro-motive force to drive a certain current through them, if their sectional area be large and their length small in proportion ; while a good conductor may offer a very high resistance, demanding the expenditure of a high E.M.F. if its sectional area be very small and its length be very great ; and it often happens that the broad, deep bed of a river, or the moist crust of the earth will offer a better conducting path for a given current than a wire stretched by its side in air, notwithstanding the fact that the specific resistance of water is so high compared with that of the metals.

The *Specific Resistance* of any given substance, then, is the proportion that the measured resistance of a given portion—say a cube of a certain size—of the substance, bears to that of a similar cube of the standard, silver.   The actual or absolute resistance of a body, or its working resistance, as it would perhaps be more appropriate to term it, is its resistance measured in actual work by the E.M.F. expended in driving the unit current through it.

It will be seen that specific resistance bears a close analogy to specific heat and to specific gravity. The facts above mentioned, as to the conducting power of large masses of imperfect conductors, are very important indeed in electrical engineering. Often one's whole calculations may be upset by the presence of a large body of moisture; and often, too, what would otherwise be very puzzling phenomena are due to this same cause—the presence in the path of an electric current of a large mass of an imperfect conductor.

The conducting power of all substances is very seriously affected by combining with other substances. Thus, the specific resistance of an alloy of two metals is higher than that of either of its components. The specific resistance of compound salts, such as sulphate of copper, is considerably higher than that of either copper or sulphuric acid. This fact is also of very great importance in electrical engineering, impure metals having a much higher resistance than pure. The presence of a very small percentage of foreign matter in copper or iron, will increase the specific resistance of the metal as much as 50 per cent. and more. The resistance of cast iron is very much greater than that of wrought iron. The resistance of a soldered joint, say two copper wires bound or twisted together and soldered, is always much greater than that of the same length of one of the wires. Hence joints are to be avoided as much as possible where large currents are used, as in electric lighting.

The fact pointed out above, also, that the resistance

of compound salts is higher than that of either of their
components, is of great importance in connection with
the working of galvanic batteries, as the substances
used in them are constantly changing, compounds being decomposed and new compounds formed.

It will be gathered also, from what has been said,
that if we know the specific resistance and the dimensions of any body, we can calculate what its actual
working resistance *ought to be;* but it must always
be borne in mind, that in our calculations there are
many things, such as joints, that we can only approximate, and we shall therefore do wisely not to work too
closely by our calculations. Always allow something
for coming up, as sailors do when hauling a rope taut.
It is usual to measure the *resistance* of any body, and
not its *conducting* power, as this lends itself more
readily to the calculations that are used in electrical
engineering, and to the measurements required for
dealing with conductors so as to produce certain results; the method of measuring actual resistances
being the E.M.F. expended when a known current
passes through the body whose resistance it is required
to know. This will be dealt with more fully later on.

The *Ohm* is the unit of resistance. It is equivalent
to the resistance offered by the following.

About 1 mile of No. 4 Copper Wire.
,, $\frac{1}{2}$,,   ,, 8   ,,   ,,
,, $\frac{1}{3}$,,   ,, 13   ,,   ,,
,, $\frac{1}{10}$,,   ,, 17   ,,   ,,

It should be mentioned, that the resistance of nearly

all substances varies with their temperature. With metals an increase of temperature gives rise to an increase of resistance. With carbon it is the reverse, an increase of temperature giving rise to a decrease of resistance. The resistance of the incandescent carbon in the arc lamp is lower than that of the carbon from which it came, and enormously less than that of the surrounding atmosphere. In the incandescent, or glow-lamp, as it is termed, the resistance of the glowing carbon filament is one-half and even less than when cold ; so that the passage of the current through the filament causes it to allow of more current passing with a given E.M.F. On the other hand, the passage of the current through a wire of a certain length, by raising its temperature, causes it to allow less current to pass with the same E.M.F.

Electric Current is the expression used to denote the fact that electricity is passing through any body or series of bodies ; and it is equally correct whether the bodies referred to are good or bad conductors,—whether they are metals, water, or even air.

An electric current passes between two points whenever the electro-motive force existing between those points is sufficient to overcome the resistance opposed to it ; and the strength of the current passing is directly proportional to the electro-motive force available, and inversely proportional to the resistance ; or, expressed in algebraical form, if C stands for current, E for E.M.F., and R for resistance, then $C = \dfrac{E}{R}$. When E is expressed in Volts, and R in Ohms, C gives the current strength in *Ampères*, the *Ampère* being the unit of current. The current strength, or the number of *Ampères* passing, is simply the expression conveying the rate of transference of electrical energy that may be going on. It may be referred back, if desired, to another term, the *Coulomb*, the unit of quantity. A current of so many Ampères means that so many Coulombs *per second* are passing at the point where the observation is made; and may be compared, for the purpose of fixing the ideas, to the rate of flow of a river or stream; it being remembered, of course, that in one case we have the flow of a material fluid and in the other that of an intangible force.[1]

The term *Coulomb* is not often used in electrical en-

[1] The direction of an electric current is always from a higher to a lower electrical potential. Thus, when an E.M.F., or difference of potential, exists between two points, one will be at a higher potential than the other, or, as it is sometimes expressed, one is + to the other; and that one is at the higher potential *from* which the current passes. Thus, of two terminals of a battery one is called +, the other —, and similarly with dynamos. In the early days of electricity it was customary to refer to the earth as zero; but since different points in the earth itself are necessarily at different potentials when a current is passing through it, this can hardly be maintained; and, as a matter of fact, we have no standard of electrical potential, nor are we able to say definitely that the electric current actually flows in a given direction; but what we do know is, that certain phenomena always follow the connection of two points between which an E.M.F. exists, and that they are always consistent with the assumption that the current passes from the point which we call positive to that which we call negative.

gineering. It is only of service when speaking ot
stored electrical energy, as in secondary batteries in
electrically charged bodies, such as clouds; or in the
conductors of long cables or telegraph lines.

Ohm's Law.—The algebraical expression above quoted
is that known as Ohm's law; and, as far as we know
at present, it is applicable to all electro-dynamic
action—that is, to all electricity in motion; whether
the current and E.M.F. used be that of telephone
apparatus, that of large numbers of electric lamps, or
that of a lightning discharge; and whether the path
open to the current be simple or complex; whether it
consist of a single conductor of one type, such as a
telegraph wire, or of a number of different types ot
conductors; and whether there be one path open to
the current or many.

It is this law, with one other which will be quoted,
which forms the basis upon which the structure of all
electrical engineering is built. It is of the greatest
service when properly used; but it is absolutely
necessary to remember its universal application.
Wherever an E.M.F. exists, it will deliver a current
through the body or bodies across which the path
may lie, exactly in accordance with the law. Thus, if
we have two main cables insulated from each other,
connected, say, by incandescent lamps, the current will
not only pass through the conducting path offered by
the lamps, but *through the insulating material;* or, in
the case of a naked wire supported by insulators,
through the film of moisture that condenses on the

insulator ; and this shows at once the importance of
making the insulation resistance in proportion to the
E.M.F. present. Thus, with a low E.M.F. a comparatively low insulation resistance may answer very
well, but would fail utterly in the presence of a high
E.M.F. This again will be further developed later
on.

It will be seen at once that Ohm's law, being an
algebraical expression, may be dealt with as any other
equation.

Thus, from $C = \dfrac{E}{R}$, we get $E = CR$.

Or, the electro-motive force existing between any two
points is equal to the product of the current passing
multiplied by the resistance.

Further, from the same equation we get $R = \dfrac{E}{C}$.

Or, the resistance of any conductor, or system of conductors, is proportional to the E.M.F. between its
extremities, divided by the current passing. This is
exceedingly useful in planning cables for electric lighting or transmission of power.

Electro-static Charge is another phenomenon that
perhaps had better be briefly referred to, though it
will not often come within the practical work of those
for whom this book is written.

Its meaning is simply as follows. It has already
been explained that electricity passes *through* bodies
which conduct. The same bodies have a certain
capacity, as it is termed, for storing it, just as bodies

have a capacity for storing heat. We know that if we wish to transmit heat through any body,—say a metal,—we have first to warm the metal, and that a certain quantity of heat will always be absorbed if a warm fluid is poured into a cold vessel; and we know, further, that different bodies absorb different quantities of heat, under different conditions. So with electricity, whenever we transmit a current of electricity through a conductor, it is necessary that the electro-static system, consisting of the conductor, its insulator, and any conductors outside the insulator, should receive a certain definite charge, or absorb a definite quantity of electricity, before the working current arrives at the other end. This is expressed technically by saying that the *condenser*, of which the conductor forms one coating, must be charged to the full potential of the current before the latter can pass on. It will probably be very rarely that the reader of these pages will require to make use of the information contained in this note; but it has a very important bearing on many forms of practical apparatus, such as long telegraph or telephone lines, submarine cables, etc. In the latter case, in fact, this phenomenon necessitated the invention of a set of specially delicate instruments to enable the received signals to be read, so much were they enfeebled after passing through a long length of cable, owing to its large electro-static capacity.

It also enters into the phenomena of lightning discharges, and of the electricity which is generated by the friction of belts over driving pulleys.

Electrolysis is the property which electric currents possess, of decomposing compound liquids through which they pass; that is to say, liquids which are formed by the combination of two or more chemical elements.

A certain electro-motive force is necessary in each case, varying with the liquid to be decomposed; and it is measured by the energy required to bring the components into their original physical condition before combination.

The combination of two or more elements sets free a certain amount of energy, which may appear as heat, electricity, or in other forms, according to the conditions present; while the decomposition of any body requires the expenditure of energy, just as an expenditure of heat energy is required to convert water into steam, and *vice versâ*.

There are usually a series of chemical reactions whenever electrolysis takes place, some of which increase the amount of energy required to maintain them, and others decrease it. Or, in other words, one set of reactions,—the combinations which take place,—set energy free and assist the current; the other set,—the decompositions,—absorb energy, and resist the current.

As also an electric current, or, more properly speaking, an electro-motive force, gives rise to decomposition, so decomposition from any other cause gives rise to electro-motive force. Thus, the strong affinities of zinc and iron for oxygen, if allowed to act, give rise to electro-motive force in the galvanic cell; while, on the other hand, the decomposition of the fluid from which the oxygen is taken, absorbs energy, and takes up a portion of that which would otherwise be available as E.M.F., the resultant E.M.F. depending upon the algebraical sum of the increments of energy set free and absorbed by these reactions.

In every form of electrolysis, whether it goes on in a galvanic cell, in an electro-plating bath supplied with an electric current from an external source, or in the numerous instances where it is always at work, though often unperceived, one rule holds good invariably. The gases, with the exception of hydrogen, are always found at the plate or point called the ANODE, where the current enters; the metals and hydrogen, at the place where the current leaves the liquid, called the CATHODE.

This property of the electric current is of great importance in many cases that are not always recognised. Thus,—wherever an existing E.M.F. can force a current through any body containing a compound fluid in suspension, it will proceed to decompose it, in accordance with the above law, and in proportion to the strength of the current,—and that is why the insulation of dynamo-machines and other apparatus breaks down occasionally without apparent reason, especially in the face of high tensions or high E.M.F.s, because Ohm's law comes in here, as elsewhere. Given a certain E.M.F. and a certain resistance opposed to it, and a certain current will pass, no matter what may be the nature of the body creating the resistance; and given a fluid, no matter how placed, with an electric current passing through it, and electrolysis will follow.

The electric current will deposit certain definite quantities of each known chemical element for each Ampère of current, as per the table below. But it must be remembered that these quantities will only be deposited provided the body has no opportunity of reforming some other combination; and in very many cases it has this opportunity, so that the rule can only be a rough guide; but, as it may be useful, it is given here.

TABLE OF ELECTRO-CHEMICAL EQUIVALENTS.

	Grammes per Coulomb.
Electro-positive :—	
Hydrogen	0·0000105
Potassium	0·0004105
Sodium	0·0002415
Gold	0·0006875
Silver	0·0011340
Copper (Cupric)	0·0003307
„ (Cuprose)	0·0006615
Mercury (Mercuric)	0·0010500
„ (Mercurose)	0·0021000
Tin (Stannic)	0·0003097
„ (Stannose)	0·0006195
Iron (Ferric)	0·0001470
„ (Ferrose)	0·0002940
Nickel	0·0003097
Zinc	0·0003412
Lead	0·0010867

Grammes per
Coulomb.

Electro-negative:—

Oxygen	0·0000840
Chlorine	0·0003727
Iodine	0·0013335
Bromine	0·0008400
Nitrogen	0·0000490

The Work Done by an electric current in passing from one point to another, through a conductor or series of conductors, is measured by the product of the E.M.F. existing between the two points, *while the current is passing*, multiplied by the current strength ; or, shortly, W, the work done $=$ E, the E.M.F., $\times$ C, the current. And, if E be in Volts and C in Ampères, the result is in *Watts*, the *Watt* being the unit of work done, and equal to $\frac{1}{746}$ of an English H. P., or $\frac{33000}{746}$ foot-lbs. It will be seen that this method of measurement of the work done by an electrical current is similar to that for all work done. Thus, the work done by a heavy body in falling from a certain height is measured by the product of its weight and the height from which it fell. Time, of course, enters into the calculation, as in all other measurements of work done, the *Watt* being equal to $\frac{33000}{746}$ foot-lbs. per minute, or $\frac{550}{746}$ foot-lbs. per second.

This formula is of great importance in connection with electric lighting, transmission of power, and other mechanical work, as it enables us to calculate the mechanical energy required to produce certain electrical energy ; and, on the other hand, having given certain electrical energy, we can calculate the amount

c

of mechanical energy we may hope to recover, after allowing for loss by heat and other conversions.

Taking the formula . . . . $W = E\,C$
and one of those given for Ohm's law . $E = C\,R$
and substituting for E its value C R in
  the first equation, we get . . . . $W = C^2 R$

This is usually written $H = C^2Rt$, and is known as Joule's law, for the heat developed by an electric current in any conductor through which it is passing ; heat, of course, being merely one form of work.

In other words, the heat developed in any conductor per second, is measured by the product of the square of the current passing, in Ampères, multiplied by the resistance offered in Ohms. C^2R is measured in Watts, just as EC was. In fact, one will stand for the other, and can be directly connected with Joule's law of heat by his well-known heat equivalent, 772 foot-lbs.

It will be seen that this law shows us how to calculate the heating effect of the current upon any conductor, whether it be the carbon filament of an incandescent lamp or a main electric lighting cable ; that is to say, the heating effect we *should* get, if there were no sources of loss, such as radiation, specific capacity for heat in the conductor, etc.

We may also obtain another formula from Joule's and Ohm's laws combined, which is sometimes of service :

$$W = E\,C. \quad C = \frac{E}{R} \quad \therefore \quad W = \frac{E^2}{R}$$

or the work done in *Watts* by an electric current is equal to the square of the E.M.F. when the *current is passing*, divided by the resistance.

This formula is of service where the current is not directly measurable, and will explain the reason that the effects of a lightning discharge are so disastrous, its E.M.F. being probably millions of volts in some cases; though, owing to the great resistance opposed to it, the current passing is small. It will be noticed that it has been stated that the E.M.F., when the current is passing, is to be taken, the reason being that the measured E.M.F., or difference of electrical potential existing between any two points, is less when a current is passing; the passage of the current over the conductors in its path, before arriving at the two points in question, taking toll as it were of the initial E.M.F. generated, in accordance with Ohm's law. This will be dealt with more fully in the chapter on electric light circuits.

CHAPTER II.

BEFORE attempting to deal with any of the forms of electrical apparatus that are in practical use, it may be as well to give some idea of that which underlies the working of all electrical apparatus, viz., the electric circuit.

The electric circuit is the path or paths by which the current passes; and it must be complete from any given point, say the positive pole of the generator, through the wires, cables, etc., forming the external circuit, back to the negative pole of the generator, and through that to the positive pole from which it started. If any portion of the circuit be wanting; that is to say, if there be any place, or any body present in the path of the current, where the available E.M.F. cannot force any current through, then no current passes in any part of the circuit; and the apparatus which should have been actuated by the current does not work. A complete electrical circuit is sometimes spoken of as a *closed circuit*, and the operation of causing the current to cease is referred to as *breaking circuit*, or *opening circuit*; and again, a

circuit in which no current is passing is sometimes
called an *open circuit.* Thus, a dynamo machine is
spoken of as being run on *open circuit,* when no work
is being done in the circuit external to the machine.
The same current,—that is, the same strength of current in *Ampères,*—passes in every part of a closed
electric circuit; so that if a body whose resistance is
comparatively high form part of the circuit, it will
weaken the current passing in every part, in accordance with Ohm's law ; and conversely, lowering the
resistance of any part of the circuit will raise the
current strength.

Though it has been stated that for all electrical
action, or rather for the working of all apparatus requiring the passage of electrical currents, a complete
electric circuit is necessary, it does not follow that
there may not be more than one electric circuit; in
fact, there may be as many of them as you like, and
they may be arranged all to emanate from one source,
or to branch out from other circuits. But no matter
how many there are, the same rules hold good, viz.,
that no current will pass in any circuit,—whether it
be one of a number of circuits, a branch from another
circuit, or a simple circuit by itself,—unless that individual circuit is complete ; and it follows, of course,
that in the case of branches from a larger circuit the
main circuit and the individual branch must be complete. Further, when a main circuit is divided into a
number of branches, technically called derivations, the
current in the main circuit divides between the branch

circuits in the inverse ratio of their resistances, the branch having the highest resistance taking the smallest current, and *vice versâ.* This again is strictly in accordance with Ohm's law.

A simple method of grasping the idea of an electric circuit, which the Author has been accustomed to place before the pupils of his firm in the early days of their articles, is the following :—

Imagine a complete ring or hoop of wire, as shown in Fig. 1 ; and that an electro-motive force arises at

Fig. 1.   Fig. 2.

some point in the ring of sufficient magnitude to generate a current. This current will go on circulating round the ring as long as the E.M.F. exists, and the wire remains intact. For simplicity, it is assumed that the ring is perfectly insulated from every other conductor, and that there are no branch circuits.

Now, suppose that we cut the ring of wire with a pair of pliers, at any convenient point, as Fig. 2. The current will no longer pass. Now, let us take a galvanic cell and connect its two poles—that is, the points at which the current can be taken from it—to the ends of the wire we have just cut, as Fig. 3. We have in the galvanic battery the source of electricity we require, and the current from it will continue to circulate through our ring of wire and our cell—which it must not be forgotten forms part of our circuit—as long as the cell continues to create an E.M.F., and there is no break in any part of the circuit, either in the wire loop or in the cell itself. But how are we to know that we have a current passing? Well, in some cases the action visible in the battery cell will tell us, but not always; and, as we shall see later on, a battery may be consuming materials when no useful work is

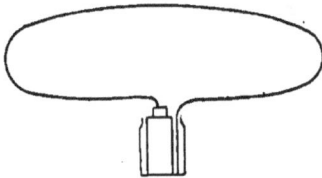

Fig. 3.

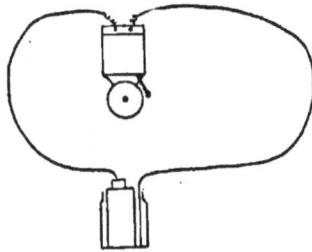

Fig. 4.

being done. Cut the wire in a second place and connect to the ends some apparatus that will denote the presence of the current furnished by one cell, against the resistance of our loop of wire; say an electric trembling bell, as Fig. 4. The current that will pass will be as the E.M.F. created by the cell, divided by the resistance of the circuit; that is, the combined resistance of the cell, the bell, and the connecting wire. If our bell is constructed to work with the current passing in our circuit, it will commence ringing, and will go on until either the battery ceases to create an

E.M.F., there is a break in the circuit as before, or the resistance of some part of the circuit—say that of the cell itself—rises sufficiently to reduce the current below the strength at which the bell will work.

The question of the increase of the resistance of the battery will be dealt with later on; at present we will only consider actual breaks in the circuit. Cut the wire in a third place, and, this time, insert a push, as shown in Fig. 5, such as are to be seen in

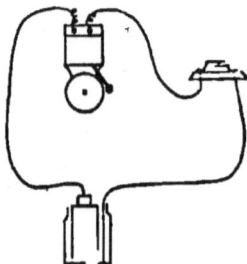

Fig. 5.

every optician or electrical apparatus dealer's window, and which consists simply of two springs mounted on a board, and so arranged that the ivory button facing them, when pressed by the thumb, brings them into metallic contact; the ends of the two wires being connected to the two springs.

If we have carried out the above experiment carefully, the bell will now ring whenever we press the button of our push; and we have control of the bell as long as we have no other break in our circuit. It is obvious, of course, that the wire connecting the push, bell, and battery may be of any convenient length, provided that the battery, wire, and bell be so proportioned that the latter will ring with the current that passes. The reader will recognise the arrangement as that of an ordinary domestic electric bell.

But now cut our wire in a fourth place, and we shall find that we have lost the control of our bell, because we have another break in the circuit besides the one at the push. (Fig. 6.)

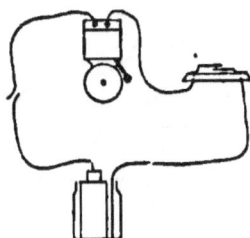

Fig. 6.

Let us suppose, also, that our wire is covered with gutta-percha, which we know has a very high resistance; and is, for most practical purposes, an insulator. Suppose that in any one of the connections we have made—to battery, bell, or push—we neglected to remove this covering before making our connection, we should find that we had no current passing; and the reason would again be, because, in accordance with Ohm's law, the resistance offered by the two thicknesses of gutta-percha was so great in proportion to the E.M.F. created by the cell, that no appreciable current could pass. Therefore, in all connections be-

tween wire and wire, or between wires and terminals
or other connecting pieces, we always first remove
whatever insulating covering may be present, and we
also scrape the wire bright, as dirt offers a higher resistance than copper. The reader must not imagine
that gutta-percha offers an impassable resistance under
all conditions ; each case works out in exact accordance
with Ohm's law. It was only because the E.M.F. of
the cell was low, that the gutta-percha barred the

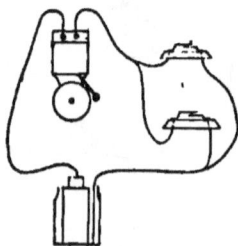

Fig. 7.

passage of the current; if a high tension generator
had been used, some current might have passed through
the covering.

Let us attach wires to the bell and battery, and
lead them to a second push, as shown in Fig. 7. We
have now a circuit with two branches, the bell and
battery being included in each, and we therefore control the action of the bell from either push, so long
as there is no other break in the circuit. If there
should be a break in that portion of the circuit between the bell and battery which is common to both

branches, we lose the control of the bell from both pushes. If there is a break, other than that at the push, in either branch outside the bell and battery, we lose the control of the bell from the push belonging to that branch.

Even a break in the circuit itself acts strictly in accordance with Ohm's law. If no current passes, it is because the resistance interposed by the break is infinitely large, compared with the E.M.F. available. If the E.M.F. were increased very much, a current might still pass, as in the case of the electric spark, the arc lamp, and in a lightning discharge; the strength of the current in each case being strictly proportional to the E.M.F. divided by the resistance; but it is always necessary to be sure that we have *all* the resistance of the circuit included in our calculation. Not infrequently the resistance of the generator itself, of the body from which the current proceeds, is of very great importance. Often, too, matter which we do not want, such as dust or dirt, interposes itself in the circuit, and before we can apply the law correctly, we must know this resistance, and it must take its place in our calculations.

It may perhaps be as well here to give the rule for determining the joint resistance of two or more derived or branch circuits. Supposing that we have a source of electricity, with a certain E.M.F. available at a certain point, from which branch out several derived circuits, as they are called. The current in each branch is determined strictly by reference to

Ohm's law; that is, the E.M.F. being the same, all we have to do is to divide it by the resistance in each case. A dynamo feeding a number of incandescent lamps is a familiar instance, and easily understood. Suppose that the E.M.F. is the same at the terminals of all the lamps; then, if the resistance of each lamp is the same, the current passing through each will be the same; but if their resistances be different, as when they are of different candle-powers, or different patterns, then we calculate the current in each case by the formula; and having obtained all the currents, add them together, and apply the formula $R = \dfrac{E}{C}$ to the whole, and we have the joint resistance of the group.

To take an example, suppose we have a dynamo, or an electric light service with 100 Volts difference of potential, and that we have lamps whose resistances when burning are respectively 400 Ohms, 200 Ohms, 100 Ohms, 50 Ohms, 10 Ohms. These lamps will take $\frac{1}{4}$ Ampère, $\frac{1}{2}$ Ampère, 1 Ampère, 2 Ampères, and 10 Ampères. The sum of these $10 + 2 + 1 + \cdot 5 + \cdot 25 = 13 \cdot 75$ Ampères, and their joint resistance $= \frac{100}{13 \cdot 75} = 7 \cdot 57$ Ohms.

The joint resistance of a number of branch circuits, whose resistance is equal, as, say, a number of incandescent lamps, all of the same pattern and of the same candle-power, is found by simply dividing the resistance of one branch by the number of branches; thus, if we have 20 lamps, each of 200 Ohms resistance, their joint resistance is equal to $\frac{200}{20} = 10$ Ohms, when they are arranged as branch circuits emanating from the same points. Of course, in actual work, you do not always have all the branches emanating from the same points, and it is therefore generally more convenient to calculate the currents than the resistances.

For two branch circuits of unequal resistance, the joint resistance may be found by multiplying their resistances together, and dividing the product by their sum. Thus, if we have two resistances 10 and 2, their joint resistance $= \frac{10 \times 2}{10 + 2} = \frac{20}{12} = 1\cdot\dot{6}$ Ohms. For any number of branch circuits greater than two, the formula becomes somewhat complicated, and the method given above, of finding the current strength in each case, will be found simpler.

This matter will be explained more fully when dealing with Electric Lighting Circuits. It may be mentioned, however, that the rule applies to such branch circuits as the insulation of telegraph or telephone lines, and to that of cables for electric lighting or transmission of power ; each unit of length, say a mile, or a hundred yards, and each pole, being considered as a separate branch. Thus, if the insulation resistance of a telegraph or telephone line,—that is to say, the resistance between it and the earth,—be 1,000,000 Ohms for 1 mile, for 1,000 miles it would only be 1,000 Ohms, and the leakage path, as will be explained later, would be very serious, as, with ordinary telegraph wire, it would be less than that of the proper conducting path.

The Earth.—In connection with the electric circuit, the EARTH, or GROUND, as the Americans term it, often plays a very important part.

It has been explained that the electric circuit consists of a path, from the generator, through the apparatus that is to be worked and that which is to work it, back to and through the generator itself. It will be seen that the apparatus to be worked—the bell, lamp, or whatever it may be—and that which is to control it—say the push, key, or switch—may be some distance apart. This requirement would necessitate two wires or conductors between the two places, unless some other path could be found that would perform the office of one wire, and carry the returning current.

The crust of the earth, a river or stream, the metals of a railway, the water and mud of a dock, the gas or water service of a town, may each form this path, and thus save one wire or cable. But it must be distinctly borne in mind that " earth," or " ground," as anything which answers this purpose is technically called, is subject to Ohm's law, exactly as the wire or cable which forms the other part of the circuit is. If the resistance of any part of the earth circuit is high, it will weaken the current, exactly as a bad joint would, or a small wire. Thus, if you make connection to the gas-pipe, and there happens to be a *good gas* joint; but, as often happens, a bad electrical joint; between one of the feed pipes and the main, you get what is technically known as " bad earth ;" or, in other words, you have a resistance in the circuit that you had not bargained for, and which must be avoided if possible. So also with reference to the rails; if the connections between the rails at the fish-plates are not electrically good, a like result will follow.

It must be borne in mind, too, that the resistance of the earth circuit must be proportionate to the current that is to pass. Thus, what would be good earth for a telephone current may be very bad earth for an electric-lighting current. Very large iron water pipes, even, might offer so much resistance at their joints as to be quite unsuitable for electric-lighting currents of even moderate strength.

Another important point in connection with earth is its variability under different physical conditions. If the crust of the earth be used, for instance, in very dry weather, or in frosty weather, the resistance may be almost infinite, as the conducting path consists of the moisture in the pores of the earth. Again, a dock with 20 feet of water in it would afford a good conducting path, whereas the dock empty might not.

It will easily be understood, too, from what has been said on the laws of resistance, that it is of importance to have a large surface in contact with the earth, or whatever may be used for a return, as the resistance offered by the earth will vary directly as the distance between the two surfaces, and inversely as their sectional area. It is also of importance that the earth-plate should be proportioned to the current that is to pass, and that it should be subject to as little chemical action as possible. Under the very best conditions there will be some resistance at the surface of the plate, where it is in contact with earth; and that resistance will generally increase, owing to chemical action. It is therefore wise, in laying down electrical apparatus, when using earth, to allow for this, and to make your earth connections very much larger than they otherwise would have to be, or than a consideration of theory simply would dictate; that is to say, the theory that would apparently apply at the time the plate was laid down.

The earth is, of course, subject to the same laws of derived circuits as other conductors; and we may expect that if a portion of the earth's crust, or a stream, gas or water service, form part of several circuits, portions of current from one circuit will occasionally find their way into other circuits; and accordingly we find this the case with telephone circuits, it being most difficult to prevent messages being heard in wires not intended for them, owing to their passage by earth. This is always more marked where earth is bad, or of high resistance.

Insulation.—Before proceeding to the practical details of the construction of the various apparatus in use, it will be advisable to devote a short space to the consideration of the most important of all points in connection with electrical engineering, viz. that of insulation. This is the rock that almost invariably wrecks the tyro in electrical matters, and more particularly even the man who, possessing but a slight knowledge of the science, fancies there is no mystery about it.

It will be remembered that it was stated at the beginning of the book that although bodies divided themselves roughly into three groups, which might be termed for convenience, conductors, semi-conductors, and insulators; yet the difference was only one of degree. All bodies conduct, all bodies offer some resistance; and whether a current passes or not, and what current passes, is decided by Ohm's law, no matter what the substance may be.

It has been already pointed out, that a long length of fine wire offers a higher resistance than a river, for instance, though the wire is classed as a good conductor, and water as a very bad one. It should also be noted that the conducting or insulating properties of *all* bodies, vary with their physical condition. Thus dry cotton is a good insulator, wet cotton a very bad one. Dry glazed earthenware or glass are good insulators; the same with a film of moisture condensed on their surface are very bad insulators.

Further, the relative values of different substances as insulators vary inversely with the E.M.F. opposed to them; and also, it must not be forgotten that the electrical resistance of the insulator *through* its substance follows the same law as to length and cross section as conductors, so that a substance which may insulate very well where only a small cross section is exposed to the E.M.F. present, may not do so if the section is large. As for instance, in the case of two covered wires touching each other outside their covering; if they touch only at one point, the resistance of the insulating material—or the insulation resistance, as it is technically called,—is high because the dimensions

D

are small ; but if the two wires are twisted, and lie together for some distance, the insulation resistance may be very much less.

A substance that will insulate perfectly in the presence of an E.M.F. of a few Volts, such as are used in telephone and electric bell work, may break down entirely under the strain, say, of 100 Volts, the E.M.F. now used in most electric light work. And again a substance that will answer for 100 Volts, may break down under the strain of 2,000 Volts, the E.M.F. which the Brush Cos. are using, and which apparently is to be used in town supply.

Again, in the choice of an insulator for any particular work, the electrolytic properties possessed by the current, and which have been already described, must not be overlooked, nor its ability to spark across short distances. A substance that may be a perfect insulator when new—if placed to separate two points or surfaces between which an E.M.F. exists,—may gradually break down from the action of the current itself. Remembering, once more, that Ohm's law holds good here as elsewhere, whatever the fraction of an Ampère of current may be which the law says shall pass ; that current *will* pass, and will do work, and will probably alter the nature of the insulating substance, silently but surely ; lowering the resistance of the insulator, till some mechanical action comes into play, such as the sharp edge of an iron plate, or a needle-point left in a casting, and either breaks the insulation down itself, or, by lessening the distance to the nearest point in the wire, provides the conditions necessary for a spark to pass, with practically the same result.

Again, the position of the points or surfaces that are to be insulated, with reference to the rest of the circuit, must be considered. Suppose a certain voltage, say 100 Volts, to be present at the ends of the wires of the exciting shunt coils of a dynamo. Let the resistance of the coils be 50 Ohms, the current passing in them will be 2 Ampères. Now, it will be apparent that we have the full 100 Volts present only between the outer ends of the coils. Between the middle and one end we have 50 Volts only, because from Ohm's law $E = CR$, $C = 2$ as before, and $R = 25$ Ohms. $\therefore E = 50$ Volts. So, if the field magnets have four legs each with 12·5 Ohms resistance, the E.M.F., or difference of potential existing between the ends of the coils on either leg, is 25 Volts only; and this is no question of theory, it may be proved by an actual test, with an instrument called a Voltmeter, to be described later, or by any of the methods described in the text-books. It is obvious that we may carry this matter as far as we like. Say that there are 25 layers of wire on each leg. Assuming that the wire is uniformly wound, each layer will have a resistance of ·5 Ohm, and as the same current passes, *viz.* 2 Ampères, the E.M.F. between any coil and the one above or below it will be 1 Volt. Further, if the layer consist of 20 turns, the resistance of each turn will be = ·025 Ohm, and the voltage between any two adjacent turns at any point = ·025 × 2 = ·05, or $\frac{1}{20}$ Volt.

Thus it will be seen that the E.M.F., or difference of potential, usually present between many points which are in close proximity in coils and other apparatus is very small indeed; and therefore as long as the voltage and insulation remain the same all goes well; but should the insulation of any part be lowered, as by wet, oil, etc., by adjacent coils coming into contact, or by the deterioration of the insulator, the normal strain is increased. Thus, suppose that a quarter of the resistance of the coils of the dynamo magnets before referred to be cut out, say by the ends of the coils of one leg having come into contact with each other, so that the current passes across this path instead of round the coils. Assuming that our E.M.F. remains the same, we have now only 37·5 Ohms opposed to the 100 Volts and $\frac{100}{37\cdot5} = 2\cdot66$ Ampères passing, instead of 2 Ampères. Our voltage will now be in each leg, $12\cdot5 \times 2\cdot66 = 33\cdot75$ V., instead of 25 Volts, and the rest in proportion.

An increase of this magnitude, on the present construction of dynamos would probably not be serious, the only thing that would happen being increased heating of the coils that were not cut out, and an alteration in the lines of force. But suppose the above figures were multiplied by 20, and the short circuiting of a coil gave rise to an increased voltage of 250 V., between the ends of the coil of one leg, then the matter might be very serious indeed.

The substances used for insulating are, silk and cotton, in places where they will not become wet, as in coils of bells, telegraph and telephone apparatus, dynamos, motors, etc., where also the space available for insulation is small, and the covering cannot easily be subject to mechanical injury. In some cases the cotton or silk is further protected by a coating of some insulating varnish, such as shellac or india-rubber; but it is necessary to avoid all possibility of chemical action between the varnish and the wire, or between the varnish and the covering.

For wires which are exposed to moisture, or that have to stand a certain amount of rough usage, india-rubber, gutta-percha and Callender's pitch compound are used, generally in combination with wrappings or plaits of cotton and tape.

For wires that run overhead, as telephone and telegraph wires, no covering is required, the air being the very best insulator obtainable, *when it is dry.* For these wires the rests are formed of highly glazed porcelain, or vitreous earthenware, made into special forms, so that the path from the wire to the iron bolt carrying the insulator, is as long as possible, and of as small a cross section as possible. Ohm's law comes in again here. With telephone and telegraph apparatus the wire is connected to ground at each end, and a branch circuit will be formed to ground from the wire by way of each insulator and its support, be it a pole or the roof of a house. It will be obvious that the resistance of a single such leakage path will be very high; and that, provided the E.M.F. be low, the leakage current must be very small indeed; and so, on short lines, it usually is. But it will also be apparent that, as the length of the line increases, the number of these leakage paths will increase also, and the leakage current may be a very serious matter.

It is this which makes the problem of telephoning over very long distances so difficult in our humid climate.

Where the E.M.F. is high, as with high tension Electric Light circuits, the leakage from even a comparatively small number of rests may be serious; but at present the loss by leakage on high-tension Electric Light circuits carried overhead is inappreciable, because the lengths of the lines are so small. It will be seen later on that the insulation resistance of these Electric Light circuits should be kept up for another reason, viz. the safety of the public.

For the insulation of apparatus and parts of apparatus used in electrical engineering, various substances are used. Hard wood, when dry, is a very good insulator for many purposes; wet wood is a very bad insulator for any purpose. Thus, a ringing key at a colliery shaft bottom, if mounted on wood, may work perfectly at first, and fail after a time, owing to the wood becoming saturated with moisture.

Vulcanite—hard, black vulcanized india-rubber—is a first-rate insulator for nearly every purpose, and it does not readily absorb moisture; but it is brittle and expensive, and moisture will condense on its surface. It is unsurpassed for small collars, knobs, etc., designed to insulate two parts of an apparatus between which a high difference of potential exists, yet which must from the construction of the apparatus be close together.

Vulcanized fibre, another substance somewhat similar to vulcanite, is of great service in many places where the brittleness and expense of vulcanite forbids its use. It is tough, and, as its name implies, of a fibrous nature. Its one drawback is, that it absorbs moisture, and then its insulation resistance diminishes very considerably. It must not therefore be used where it will be exposed to damp or oil, except where only a low E.M.F. is present.

The flexible fibre is not a good insulator.

Mica and asbestos are also used for insulation, chiefly on account of their non-combustible properties ; but they can only be used in certain cases. Asbestos, when worked into mill-board, answers very well for many purposes, such as the insulation of commutator sections, coils of magnets, etc., always provided it can be arranged that the sheet shall not be torn ; but it is not a perfect insulator.

The use of mica is more limited. Owing to its peculiar laminated character, you can have a plate of mica of a certain size as thin as you like, but it must be a plate of one thickness all through. It is not workable to section like other substances. It is somewhat brittle too, and it is doubtful if its insulating properties are as high as some people think. Slate and porcelain are now being used for the bases of Electric Light switches and fuzes ; but the former is not a good insulator, and will not answer at all for high E.M.F.s ; the latter has the disadvantage that it is difficult to work and is easily broken.

Once more, it must not be forgotten, that in the use of all these substances, Ohm's law is the sole arbiter of the fittest, coupled, of course, with the law of dimensions and resistance. Thus, it may be quite practicable to use a comparatively poor insulator in the presence of a low E.M.F., especially if the insulation path is or can be made long and of small cross section, where it would not be if the conditions are reversed.

Induction.—A series of phenomena in connection with electricity that have a very important bearing upon the working of all electrical apparatus, are what are known as electrical induction, or electrical action at a distance.

It has been explained how electrical currents pass through conductors where continuity exists—where continuity does not exist, another series of actions takes place—induced electro-static charges are formed, and induced currents are generated.

When an electro-static charge is held upon a conductor, completely isolated from other conductors, a charge of an opposite name is *induced* upon all other conductors in the neighbourhood, that are not isolated.

Again, when a current of electricity passes round a piece of iron, magnetism is *induced* in the latter, and will be rendered visible on closing the magnetic circuit.

When a permanent steel magnet is brought near a piece of iron or steel, magnetism is *induced* in the latter, provided it lies in the path of the magnetic circuit.

But the most important phenomena of all are magneto-electric induction, and the induction of currents upon each other.

It a permanent or an electro-magnet be brought into the neighbourhood of a conductor, so that the latter lies at right angles, or nearly so, to the path of the magnetic circuit, or to the lines of force as it is usually expressed in the text-books, an E.M.F. is generated in the conductor, which will give rise to a current, if a path be open for it ; and this generation takes place as long as the motion continues, or as long as an alteration in the field in the neighbourhood of the conductor is going on. Thus, suddenly exciting an electro-magnet whose magnetic circuit crosses a conductor, will generate an E.M.F. in the latter. Suddenly causing an electro-magnet to lose its magnetism will have a similar effect ; but the E.M.F. in the former case will be in the reverse direction to that in the latter ; that is, it will tend to produce a current in the opposite direction through the electric conductor. Varying the strength of an electro-magnet will have the same effects, though in a minor degree, as suddenly magnetizing it or causing it to lose its magnetism.

Upon the phenomena of induction the dynamo-electric machine, the induction coil, and the transformer have been reared.

The property of inducing currents also extends to wires in the neighbourhood of other wires. If, for instance, a second wire be wrapped round an electro-magnet, in addition to the exciting wire; each time that the exciting circuit is closed, an E.M.F. will be generated in the second wire ; and each time the exciting circuit is broken, an E.M.F. will be generated in the second or secondary wire, as it is usually termed, the exciting wire being called the primary; but the direction of the E.M.F. generated in the two cases will be opposite to each other.

A variation in the strength of the current passing in the primary or exciting wire is also followed by a generation of E.M.F. in the secondary, though in a minor degree.

The directions of the secondary E.M.F.s are always such as to resist the action of the primary current. Thus, the current which passes in the secondary when the primary circuit is closed, is in the opposite direction to that passing in the primary ; that which is generated in the secondary when the primary circuit is broken, is in the *same* direction as the current that was passing in the former.

Similarly, weakening the primary current generates a current in the secondary in the same direction as that which is passing in itself. Strengthening the primary has the reverse effect.

It is not necessary even for two wires to be together on an electro-magnet for induction to take place. Suddenly making or breaking the exciting circuit of an electro-magnet, generates opposing E.M.F.s within the coils of the exciting wire itself; that when it is made opposing the current, and that when it is broken assisting it. The latter has been known as the *extra* current, having been so named by Faraday, to whom we are indebted for so many researches upon electromagnetic induction. It is the *extra* current that gives such a smart and often fatal shock, when the circuit of a high tension Electric Light machine is broken; the coils on the field magnets of the dynamo generating, by induction, a very much higher E.M.F. than the working E.M.F. of the machine.

The secondary E.M.F. generated in all these cases depends upon the primary E.M.F., the number of coils taking part in the induction; or, what amounts to the same thing, the lengths of the wires exposed to induction; upon the speed of motion where one or both bodies move; and inversely upon the distance between the exciting and secondary apparatus.

It must be remembered, however, that in all these cases, induction only takes place while motion is proceeding, or changes of magnetism are taking place, and the currents generated are therefore usually only of very short duration.

But it is not even necessary that iron should be present, for induction to take place. If two wires be near each other and parallel; when a current passes in one, an E.M.F. is induced in the other at the moment the first starts and at the moment the first ceases; and these two are in opposite directions, and obey the same rule as before; viz., the secondary current is in the opposite direction to the primary, when the latter commences, and in the same direction as the primary when the latter ceases.

A variation in the current passing in one wire also gives rise to induced E.M.F.s in the other, just as in the case of the electro-magnet, with two wires wound on it.

The induction in the case of two wires also follows the same rule as to distance apart, and to the lengths of wire exposed to induction. The induction, for instance, between two telephone or telegraph wires running parallel for several miles, and within a few inches of each other, as on ordinary telegraph poles, will be very great; while that between wires separated by the width of a street, or only running together for a short distance, may be inappreciable.

The reason for E.M.F.s being generated in conductors under the conditions named, is evidently due to their passing through the magnetic circuit, or, as it is usually termed, cutting the lines of force. Where an electro-magnet is excited, a magnetic field, a magnetic circuit, or magnetic lines of force, whichever term be preferred, is created; and the secondary wire lying in their path, an E.M.F. is generated in it. The converse takes place when the primary circuit is broken; and this applies to all cases of electro-magnetic induction. A magnetic circuit is either created, broken, or varied; and in each conductor lying in the path of the circuit, an E.M.F. is generated.

From the above it naturally follows, that in conductors which run parallel to the directions of lines of force, no E.M.F. is generated; and so it happens that

no induction takes place between two wires crossing -
each other at right angles.

For effective induction, where the secondary current
is to be used apart from the primary ; as in induction
coils used for experimental purposes, for telephone
transmitters, for the transformers now being introduced into electric lighting ; the primary and secondary coils must be insulated from each other, and the
insulation should be proportionate to the E.M.F.s
generated. With induction coils, by wrapping a long
length of secondary wire round a shorter length of
primary, it is possible to generate very high E.M.F.s
in the former, using only very low E.M.F.s in the
latter ; each coil of the secondary that is brought
within the influence of the lines of force, adding to
the E.M.F.s generated. Where high E.M.F.s are
generated, and used for the purposes of Electric Light
distribution, being converted into low E.M.F.s at the
point of consumption, great care must be taken to insulate fully between the coils, or the inductive action
will be destroyed to a large extent ; and may give rise
to serious accidents.

But though, where it is desired to use electrical induction as a servant, it is necessary to insulate the
two coils from each other, induction will take place
nevertheless without any insulation. One case has been
given, where the coils of an electro-magnet generate
an E.M.F. by acting inductively on each other ; coils
even are not necessary for induction. But it will take
place within a wire itself.

Each portion of a conductor acts as a separate wire. Thus, a copper rod half an inch in diameter, may be taken to consist of a number of small wires all grouped round a centre; and as apparently the first action of a current is mainly confined to the surface of a conductor, induction takes place between the outer wires, so to speak, and the inner, giving rise to many puzzling phenomena in connection with dynamo construction, and with lightning discharges; about which something will probably be said later on, Prof. Oliver Lodge's recent investigations having considerably modified our ideas with regard to the latter.

CHAPTER III.

MAGNETISM and Electricity are very closely allied, each being more easily converted into the other than into any other form of energy. In fact, as will be seen later, the presence of an electric current implies magnetism, or, as it is termed, a magnetic field around the conductor through which it is passing; while any change of the magnetism of any body or system,—or, as it is technically expressed, any change in the strength of the magnetic field,—immediately gives rise to electric currents.

Like electricity also, we know very little of its nature, though rather more than we do of the sister science. We know, for instance, that an iron bar when magnetized is longer than when not magnetized; and from various researches, principally by Professor David Hughes, we gather that the act of magnetization consists in an attempt by the molecules of the body to turn on their axes, and to place themselves in line with the direction of the magnetizing force. All bodies conduct magnetism, some with greater facility than others; but of all bodies, iron and its compounds, steel and cast iron, alone exhibit appreciably the magnetic properties, which are :—

47

1. The property of pointing, when freely suspended, to those spots on the earth's surface known as its magnetic North and South Poles,—one end, or one pole, as it is termed, turning always towards the magnetic North Pole, and the opposite one to the magnetic South Pole.

2. The property of repelling the pole that,—if both were freely suspended beyond the influence of magnets other than the earth,—would point in the same direction, and of attracting the opposite pole. Thus, two north-seeking ends of a magnet; or, as they are called, two north poles, repel each other, and two south poles repel each other; while a north and south pole attract each other.

3. The property possessed by a magnetized body of inducing magnetism in other bodies not in contact with it, and in such a manner as to cause motion between the two, if either is free to move, or if the inducing force is sufficiently powerful to overcome the mechanical resistance to motion.

As with electricity also, we have a magnetic circuit, which consists of a closed ring of attractions, and whose resistance to magnetization varies with the substance and with the dimensions of the magnetic conductor. Thus, a long, thin bar offers a greater resistance to magnetization than a short, thick one; and, stated shortly, the magnetic resistance offered by any body varies directly as its length in the direction of the magnetic circuit, and inversely as its cross section.

Steel offers a greater resistance to magnetization

than pure wrought iron, as also does cast iron; and
all three have the property of retaining their magnetism to a certain degree, when once they have been
magnetized. As might be expected, upon the theory
that the operation of magnetizing consists in a twisting
of the molecules, those bodies, as steel and cast iron,
which offer most resistance to the inducing force, as
it is termed, retain more of their magnetism, and for
longer, than wrought iron. With very pure iron, such
as Swedish, Farnley, or Lowmoor, it is very difficult to
find any traces of magnetism when no inducing force is
present, while either will give a very high return for
a given magnetizing power that is applied. With steel
or cast iron, on the contrary, it is by no means easy to
induce them to accept of magnetization; but when once
it has been accomplished, they retain a very large percentage for some considerable time.

These properties are of very great importance in the
construction of electrical and electro-magnetic apparatus. Thus, the needles of compasses and miners' dials
are always made of steel, their property of pointing
north and south being all the service that is required
of them. In any apparatus where a selective action is
required (or where an attractive power is required in
a light, portable form, as in the magneto-telephone
receiver), a piece of magnetized steel, or a steel magnet,
as it is called, is used.

Where it is required to control the magnetic effect
for the purpose of producing motion, pure wrought iron
is generally used, for the double reason, that a smaller

E

weight will answer than with cast iron for the same work, and that it responds more readily to the magnetizing influence,—usually an electric current,—taking up and losing its magnetism readily at the will of the operator.

In dynamo electric machines,—where parts of the apparatus are required to retain their magnetism in the same sense, that is, with the same polarity, as it is termed, as long as the apparatus is working,—those parts may be constructed from wrought iron or cast iron, according to the fancy of the designer; but he will have to provide a heavier weight of the latter, to do the same work.

Those parts containing iron which are in motion and continually changing the direction of their magnetization, as the iron core of the armature, are always made from the purest wrought iron obtainable.

It has been stated that there is a magnetic circuit of attractions, or a continuous path for the magnetism, just as there is a continuous path for an electric current; but with this difference, that magnetism always passes, no matter how great the resistance, and the effect of the magnetic resistance is simply one of degree.

Air, for instance, has an enormously greater resistance than iron, some 1400 times, according to a recent investigator, Mr. Kapp, to whom dynamo manufacturers are very much indebted for his able researches, and above all for his exposition of the law of magnetic resistance. Yet, if no other path be open, the magnetic influence will pass through air, imperceptibly, of course, until some body, such as a piece of iron, able to denote its presence, is placed within its influence.

Perhaps the method adopted for explaining the working of the electric circuit may be of service here. Imagine a ring of iron or steel, not covered. Iron or steel, because, as has already been explained, these substances conduct magnetism better than any others, so far as we know at present; uncovered, because we know of no substance that will act as an insulator for magnetism in the same sense that india-rubber or gutta-percha does for an electric current. The bodies which offer the highest resistance do conduct magnetism, even under moderate exciting power, to a very appreciable degree.

If we apply to any part of our ring an exciting power, such as an electric current passing in a wire wrapped round the iron, moderate in proportion to its resistance, —that is, which will develop a moderate degree of magnetization in opposition to the resistance of the ring,— we shall find scarcely any traces of magnetism anywhere outside of the ring, though we can show, by suitable apparatus, as will be seen when we come to deal with transformers, that the magnetism is there, and is a perfectly measurable quantity. It has only passed by way of the ring, just as the electric current passes by way of a wire, because, with the force available, that is for practical purposes the only path open to it. Now let us cut out, say one-sixteenth of the ring; and we shall find that we have created a totally different set of conditions, giving rise to totally different phenomena.

First, then, the piece we have cut out—if free to move, and placed at such a distance that the magnetic power is sufficient to overcome the friction, mechanical inertia, etc., present—will move back exactly into its old place, or as nearly as it can, when the exciting power is applied.

We shall find also that, if our ring is made of steel, it has retained a certain power of attracting the piece that we have cut out, after the exciting power has been removed. If our ring is of wrought iron, it will pull the piece up sharply when the exciting power is applied; while, when the current is broken, it can easily be removed, and may even fall off itself if it is heavy in proportion to the attracting power. If we pursue the matter a little further, we shall discover another and very important phenomenon. It has just been remarked that, in the case of wrought iron, the piece of the ring we had cut out might detach itself under certain conditions. Let us find the conditions under which it will not detach itself. Let the piece we have cut out of the iron ring be replaced by a piece fitting very exactly, the four surfaces being planed true to each other. Now, we shall find that we have a slightly increased holding power when the ring is excited; and further, that after the exciting power has been removed, it will still require the expenditure of a certain amount of energy to pull it away. The reason is, that while the magnetic circuit is complete, a certain amount of magnetism remains in the iron even after the exciting power has been removed. Forcibly pulling away the armature, or keeper,—as the piece of iron employed to close the magnetic circuit is usually called,—dissipates this residual magnetism; and so, if we replace the keeper ever so truly and carefully in its place, we shall not require the same effort to remove it as before. This phenomenon would form a very serious drawback to the working of many forms of electromagnetic apparatus, if there were no means of getting over it. Fortunately there are. It is sufficient if we break the iron magnetic circuit by a film of air, a layer of paper, or, more conveniently, a thin plate of brass, to get rid of what would otherwise be the troublesome effects of residual magnetism.

We shall also find another phenomenon arise, from our having broken our ring by taking out a sixteenth part as described. It has been explained that the magnetic influence always passes, no matter what may be the medium; the effect of the insertion of a resisting medium being merely the reduction of the strength of the magnetism; and it was also explained that though magnetism passed invisibly, it could be rendered evident by the presence of iron. This may be shown in a very striking manner. If we take our ring minus its sixteenth, place a sheet of paper or glass over it, and sprinkle iron filings over the aperture, we shall find these filings, when free to move under the influence of the exciting power, arranging themselves in regular order from one end of our ring to the other, across the break; and if we investigate the matter further, we shall find that the reason these iron filings arrange themselves in this way is, because each has become a small magnet, under the influence of the exciting power of our ring, and each places itself in accordance with the two laws before stated, viz. :—

1. That, it being a magnet, having a N. pole, that should point to the S. pole of the large magnet, and the S. pole of the filing to the N. pole of the large magnet.

2. That the N. poles of the filings repel the N. poles of other filings, and the S. poles the S. poles; and the joint operation of these two laws produces the curves shown, which Faraday termed the lines of force, or the direction in which the magnetic force is manifested.

As a ring would not usually be a convenient form for a magnet that is to be used in electro-magnetic apparatus, the horse-shoe form is generally adopted. For permanent steel magnets, a strip of steel is bent into the form shown in Fig. 8, and its magnetic

Fig. 8.—Horse-shoe Steel Magnet.

circuit is closed by another strip of steel or iron, or by some portion of the apparatus of which it is to form a part. For electro-magnets, the horse-shoe form is also generally adopted, but it is made in four pieces instead of two ; viz.: two limbs upon which the wire that is to carry the exciting current is wound ; a yoke, or back piece to complete the magnetic circuit on that side ; and the armature, facing the poles, and moving in accordance with the will of the operator who controls the electric current, to complete the magnetic circuit on the other side, as Fig. 9.

Lines of Force.—Before going further, it will be as well to deal with Faraday's very beautiful, but somewhat puzzling, conception, the lines of force.

Lines of force bear the same relation to magnetism

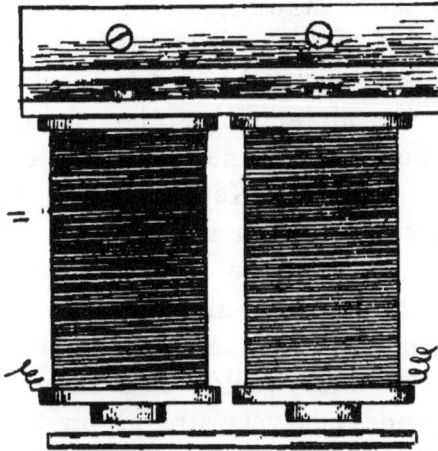

Fig. 9.—Four-piece Electro-magnet with Armature.

that current does to electricity. The production of lines of force is the result of the work done by a given magnetic exciting power, in opposition to the magnetic resistance opposed to it; just as an electric current is the result of a given E.M.F. acting in opposition to a given electrical resistance. Moreover, the number of lines of force passing at any point is a measure of the strength of the magnetism at that point, just as the number of Ampères passing in any electric circuit, or part of a circuit, is the measure of the strength of the electric current passing. As in an electric circuit also, the current strength is the same in every part of the circuit, so too in the magnetic circuit, the number of lines of force passing is the same in every part; with the proviso, as with the electric circuit, that if two or more paths are open to the lines of force, they will divide between those paths or branch circuits in the inverse ratio of the resistance of the several paths. Thus, if there be two paths open to the lines of force, one through the air and the other through iron, the dimensions being the same, they will divide in the ratio, according to Mr. Kapp's figures, of 1440 : 1. One part passing through the air and 1440 through the iron. It can easily be seen, however, that if the air path should be short and of large cross section, while the iron path was the reverse, an appreciable portion of the lines would pass through the air. This point comes out very strongly in the matter of designing dynamo machines ; what is known as the leakage path being in some cases of comparatively low resistance, owing to the form of the machine.

As with Ampères also, lines of force are definite measurable quantities. Dynamo manufacturers calculate how many lines of force they have passing into an armature, and what the section of the iron should be to accommodate them, just as they calculate the number of Ampères required in a given case, and the section of conductor required to accommodate them. But we have no convenient quantities as yet, like the Volt and the Ohm and the Ampère, that we can refer to as analogous to the foot-pound in mechanics. We have no familiar name representing so many lines, though probably it may not be long before we have one. In order to render the subject clear, therefore, we can only refer back to the foundations of all these units, those of force, mass, and time. The unit line is that force which will move the unit body, the gramme, over unit distance, the centimetre, in unit time, the second. Therefore, when we say that there are so many lines of force passing into the armature of a given dynamo, we mean that we have the power present within the armature to do that number of centimetre-grammes of work in unit time, under the influence of the magnetism created in the machine ; and, as engineers well know, these quantities are directly convertible into the more familiar foot-pounds, and H.P.

Lines of force, then, show the direction in which the magnetism present will cause any free magnetizable body to move; and the number of them at any point, referred to unit quantities as detailed above, show the force of magnetism available there, or, as it is termed, the strength of the magnetic field.

It must not be imagined that magnetism and lines of force represent continuous motion of the molecules of the body through which the magnetism passes; or that we should be justified in calling what we now know as lines of force, a magnetic current. They are nothing of the kind, nor are they analogous to an electric current, except in so far as has been described, and for two reasons. First, because magnetism is an influencing or inducing force, like gravity, and not a moving force like heat or electricity ; and secondly, we *have* a magnetic current, viz., the motion that takes place in the magnetic circuit, among the molecules of the bodies through which the magnetic influence passes, at the moment the exciting power is applied, and at the moment when it ceases.

It will be easily understood from what has passed, that as the lines of force passing through air, say from

Fig. 10.

pole to pole of a magnet, can radiate in all directions, unless their path is shaped for them by the introduction of some piece of iron, such as the armature, the force exerted varies, as in all similar cases, inversely as the square of the distance from the poles. It is doubtful, however, if this law holds good in a sufficient number of cases to be of any value, as it is evident that it must be subject to modification by every change in the conditions present.

It has been mentioned, that the existence of an electric current passing through a conductor, implies a magnetic field around the conductor. Thus, if a current be passing in a wire, lines of force are created in concentric circles around it (Fig. 10). That is to say, a small magnet such as a magnetized steel filing, if free to move, would place itself tangential to a circle of which the wire formed the centre, and simple iron filings take up positions in concentric circles round the wire, just as they do in curves across the poles of a magnet. As before, the lines of force show not only

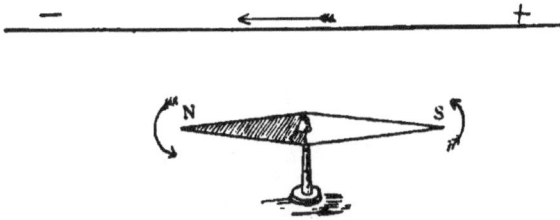

Fig. 11.

the direction of the magnetic influence, but their number or density shows the strength of the field surrounding the wire; and this again is proportional to the strength of the current passing, and inversely to the distance from the conductor.

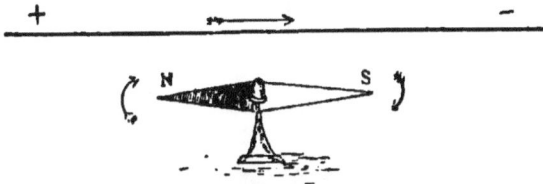

Fig. 12.

The first effect of this property of electric currents that became of any practical value, was the power which it gave of deflecting a magnetic needle out of the position we have seen it assumes, pointing parallel

ET

with the line of the earth's N. and S. poles.   Thus, if a
magnetized compass needle be suspended under a wire,
it will be found that when a current passes in the wire,
the needle will be deflected, the N. pole going to the

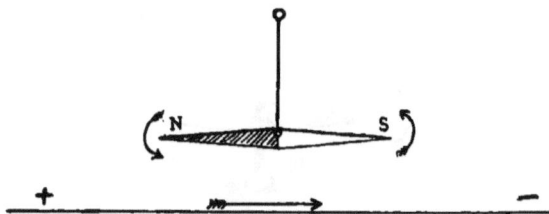

Fig. 13.

left when the current passes from S. to N. (Fig. 11)
over it, to the *right* when the current passes from
N. to S. (Fig. 12) *over* it, to the *left* when the current
passes from N. to S. *under* it (Fig. 13), to the *right*
when the current passes from S. to N. *under* it (Fig.

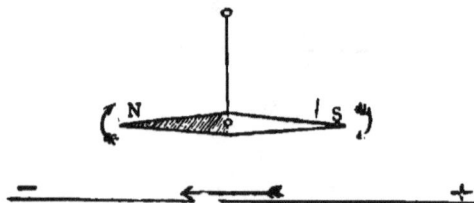

Fig. 14.

14).   Thus, a current passing from N. to S. *over* the
needle and coming back from S. to N. (Fig. 15) *under*
it, will deflect the needle in the same direction, the
N. pole turning to the right; while a current passing

from N. to S. under, and coming back from S. to N.
over, deflects the N. pole to the left; and on this
principle a large number of electrical measuring instruments are made; an insulated wire being passed

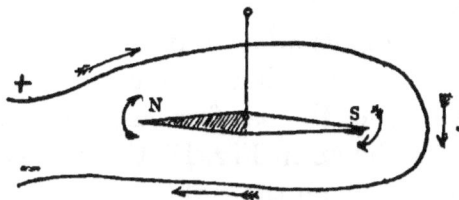

Fig. 15.   ·

continuously round the needle in the same direction.
It will be obvious that such an apparatus enables us
to discover the direction of any current passing through
the wire coils; and further, that, as the influence of

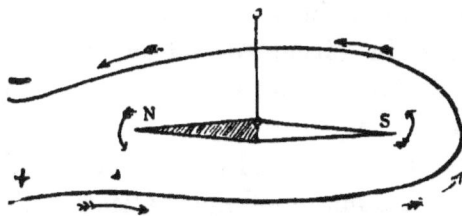

Fig. 16.

the current upon any given needle will vary with the
number of turns, and with the current passing through
the coils, we have to our hand the means of measuring
the strength of the currents we are using.   This will

be dealt with fully when measuring instruments are under consideration.

Electro-Magnetism.—The law which has been enunciated, which rules the deflection of a magnetic needle when suspended under or over a wire, or inside a wire coil, holds good equally if, in place of a movable magnetic needle, we have a bar of iron or steel, surrounded by a coil of wire in which a current is passing. If the current passes over the bar from us, and returns under towards us, as shown in Fig. 17, the N. pole is on the left, and the S. pole on the right.

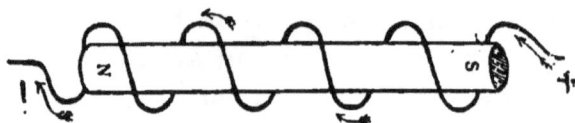

Fig. 17.

From this it will be seen that we are able to magnetize a bar in either direction, with any given wrapping of wire, by simply reversing the direction of the current passing in the wire.

It will be understood also, that the number of lines of force, or the amount of magnetization developed in any bar, will vary in accordance with the law we have stated, viz.:—it will depend directly on the exciting power, and inversely on the magnetic resistance opposed to it.

The exciting power, again, varies directly as the strength of the current passing in the wire, and as the number of times it passes round the bar, or, shortly, as the Ampère-turns. And it will follow immediately from these two, that with a short thick magnet we require less exciting power, fewer Ampère-turns, than with a long thin one; since, other things being the same, the magnetic resistance of the former is less than that of the latter. The resistance of each leg, taking the horse-shoe form, is less; and also the resistance of the back, or yoke, of the armature and of the air space between the poles and the latter, because the cross section will be necessarily larger.

This is in direct contradiction to the ideas that prevailed in the early days of electricity, and to those which are to be found in some of the older text-books. In those days it was thought that you gained power by increased length. The error was probably due to the actual fact that you may gain power by using a long electro-magnet, but from a totally different cause, viz., the increase of the exciting power. Electro-magnets are often constructed by taking a piece of iron and wrapping on it as much wire as you can. As each succeeding layer of wire is farther and farther from the iron, its exciting power becomes less in proportion, whilst its electrical resistance becomes greater in proportion to its length, so that you soon reach a point at which there is no advantage gained by adding more wire. If you take a longer bar, you increase your magnetic resistance, but not so much as to balance the increased exciting power that you gain from the additional wrapping space.

Steel is never used for electro-magnets, except in special cases, and cast iron only in the case of the field magnets of dynamo, where the direction of magnetization does not change. For all other purposes, wrought iron of the very purest quality obtainable is used; the dimensions are either calculated or determined by experiment, or calculated from *data* furnished by experiment; and the exciting power is an electric current passing in a cotton or silk-covered wire, wrapped round and round each limb of the magnet.

CHAPTER IV.

THE theory of the galvanic cell has already been sketched under the head of Electrolysis. It may be briefly summarized for practical purposes thus. Whenever we have two dissimilar conductors present in a liquid, we have a galvanic cell; an apparatus that will generate an electric current, if a circuit be provided for it.

The galvanic cell usually consists of two metals, or a metal and carbon, immersed in some electrolyte; that is, some liquid which the electric current can decompose; the whole being contained in a vessel of glass, earthenware, or other suitable substance. A galvanic battery is simply a number of galvanic cells connected together. The containing vessel is not a necessary part of the cell. Such an arrangement as an iron staple, holding a copper wire, moisture being present, forms a galvanic cell; so also does the combination of a metal with its oxide adhering to it, when moisture is present.

The term is susceptible of even wider significance, and may be taken to include any combination of bodies giving rise to an E.M.F. by means of chemical reactions. Thus, two gases, or a gas and a metal give rise to E.M.F.; and any arrangement generating a current, in which they are included, would form a galvanic cell. It must further be understood, that the different apparatus which will be described as forming parts of a galvanic cell or battery, are really devices for completing the electric circuit and maintaining it in its normal condition, so that a current may pass through it in accordance with Ohm's law. It must always be remembered that the generator, whatever it be, forms part of the electric circuit, and that any disconnection occurring within it, such as a broken plate, or no liquid in the cell, will prevent the passage of the current, just as if a wire were broken outside. Further, any increase or decrease in the resistance of the cell or other generator, the internal resistance as it is called, increases or decreases the resistance of the whole circuit, or of any portion of the circuit in which the generator is included, and weakens or strengthens the current passing, exactly in accordance with Ohm's law.

In the case of the galvanic generator, also, we frequently have more than one E.M.F. arising within the cell, owing to the secondary reactions that are necessary or unavoidable; and, as has already been explained in connection with Electrolysis, the E.M.F. obtainable at the terminals of the cell will be the result of the algebraical sum of all the E.M.F.s present within the cell. It therefore follows that any chemical reaction which gives rise to a fresh E.M.F., either assisting or opposing the initial E.M.F. at the generating plate, modifies the final E.M.F. obtainable at the terminals to that extent.

One other point should also be noted. The internal resistance of each cell takes toll of the E.M.F. generated, just as any other part of the circuit, so that the E.M.F. at the terminals of any galvanic cell or battery of cells, will be higher when there is no current than when any given current is passing, by the E.M.F. expended in driving that current through the cell, in opposition to its internal resistance, measured by the formula $E = CR$, as before.

The most favourable combination for a battery would consist of two conductors immersed in a liquid, one having what is called a strong chemical affinity for one component of the liquid, while the other conductor had an affinity for the other components. Unfortunately such a combination is never obtainable; and the nearest approach to it is when one conductor has a strong chemical affinity for one component of the liquid, while the other component, or components, are caused to combine with a third substance placed near the other conductor for the purpose; since, unless the other components are disposed of, they form, with the conductor which is not acted upon, one of the opposing E.M.F.s already referred to.

Thus the simplest and earliest form of battery, consisting of zinc and copper plates immersed in dilute sulphuric acid, is utterly useless for practical purposes, because the hydrogen which is liberated on the copper plate forms with it an E.M.F. opposing that generated by the combination of the zinc with the other portion of the sulphuric acid and water.

All known galvanic batteries that are in practical use, with one exception, have zinc as the generating or consuming plate, the one that furnishes the power, just as coal does in a furnace. Nearly all batteries also use carbon as the plate which is not acted upon, and which is called the collecting plate. A few forms use copper, and one platinum.

One form of battery uses iron for the generating plate, but it is confined to electric lighting, and has done very little practical work up to the present, so far as the Author is cognisant.

The principal variations in the different forms of battery in use at the present day, are in the exciting liquid, as the electrolyte in which the conductors are immersed is called; and in the means provided for disposing of the hydrogen gas—which always appears at the copper or carbon plate—as fast as it is liberated.

The number of different exciting liquids in use is also very small. They are: sulphuric acid, which is used in some forms of the Daniell battery, and in all bichromate and nitric acid batteries; sulphate of zinc, which is also used in Daniell's batteries; chloride of zinc, used in the Upward and the chloride of silver cells; sal-ammoniac, which is used in the Le Clanché and Sulphur Sal-ammoniac cells.

The substances used for getting rid of, neutralizing, or oxidizing the hydrogen gas, are :—

Sulphate of copper in the Daniell's battery, the

hydrogen displacing the copper from the solution, and the latter being deposited.

Bichromate of potash, chromic acid, and nitric acid, which are used sometimes separately, and sometimes in combination, in the bichromate, Grove's, and Bunsen's batteries, the hydrogen being neutralized by a rather complicated series of reactions taking place within the cell, resulting in the formation of water, chrome alum, and other salts.

Chloride of silver, which is used in Dr. Delarue's battery, and in which the silver is deposited.

Oxide of manganese, in the Le Clanché battery, the oxide being reduced to the sesqui-oxide, so that some of the oxygen previously held by it in combination is set free to combine with the hydrogen gas, and form water.

Sulphur in the Author's battery, the hydrogen combining directly with it, in the proportion of two atoms of H to one of S, and sulphuretted hydrogen being formed.

Chlorine gas in the Upward, the hydrogen gas forming hydrochloric acid with it.

In the Schanschieff cell, a solution of a mercurial salt forms at once the exciting and the neutralizing liquid.

Hydrogen gas, when newly liberated from combination by the action of the electric current, is in a very powerful state, and from its well-known strong affinity for oxygen, is able to split up many compounds containing it.

In no battery are the chemical reactions confined
to the simple oxidation of the zinc and the combination with the hydrogen. The compounds formed
in each case react upon each other, both chemically
and electrically, upon the primary compounds from
which they were formed, and upon the conductors
themselves. Thus the zinc oxide, or zinc sulphate,
or zinc chloride, produced at the initial stage of the
generation of the current, form, with the zinc plate
and the liquid in which both are immersed, a secondary
galvanic battery, whose circuit is always closed ; and
it is therefore very important that the liquid electrolyte, in addition to containing an element having a
strong affinity for zinc, should also dissolve the compound which the zinc forms, as fast as it is produced.
Failure to perform this office introduces within the
battery, at the generating plate, a resistance,—or its
equivalent, an opposing E.M.F., or both, — and so
weakens the current generated under given conditions,
in accordance with Ohm's law.

So, too, in the Le Clanché cell, the zinc chloride
forms with the zinc and the sal-ammoniac what are
called secondary salts, which are fortunately soluble
in a saturated solution of sal-ammoniac, and therefore
do no harm, so long as the liquid is saturated.

Again, the sulphuretted hydrogen generated in the
Author's Sulphur Sal-ammoniac battery, does not come
away, but immediately forms with the zinc chloride
a new set of chemical reactions, resulting in the reformation of sal-ammoniac, the original exciting liquid.

Another important feature, therefore, in the success of galvanic batteries is, that the exciting liquids shall dissolve, or dispose of all secondary salts that are formed, and not allow them either to introduce a resistance or to generate an opposing E.M.F.

In all galvanic batteries the outer containing vessel is either of glass, stoneware, vulcanite, or prepared wood. The latter is very seldom used, and vulcanite only when a special object, such as portability, is in view. In most cases glass or earthenware is used. Ordinary jam-pot ware will not answer, as the acids used in some batteries soon destroy the glaze and attack the ware. The glass and the earthenware should be of the very best quality.

Nearly all batteries in practical use have a porous division to separate the exciting liquid from that used to neutralize the hydrogen. In the cases of the Le Clanché, the Upward, and the Author's battery, the porous division is merely a containing vessel for the substance surrounding the carbon plate. In one form of Le Clanché, the agglomerate, it has been dispensed with. These porous divisions are usually made of fine clay. In a few forms of galvanic battery, that need not be dealt with here, the action of gravity is called into play, the heavier substance being placed at the bottom of the jar, and the lighter allowed to float on it.

The porous division only *retards* the mixing of the liquids. When it is used to separate them, it does not prevent their mixing with time; and that is one of the greatest difficulties attending the use of galvanic batteries where the current is required to be maintained at a constant strength. And the difficulty is increased in some cases by the action of the current itself, which always tends to carry the liquid it passes through in the direction in which it is going itself. Thus, the liquid in the carbon cell will always rise, after a few hours' work, from the passage into it of a portion of the liquid in the zinc cell.

Where the porous cell merely acts as a containing vessel, as in the Le Clanché and Sulphur cells, the liquid is required to work through, as otherwise there would be no passage for the current.

The porous cells that are used in batteries vary somewhat, according to the materials employed in the battery. Thus, for bichromate and nitric acid batteries, the porous cells are best a trifle hard—not too hard, or they will offer too high a resistance.

For Le Clanché and the Author's Sulphur battery, a softer porous cell is better. By hard and soft porous cells is meant, of course, hard-baked and soft-baked, the former not being so porous as the latter. The reason that harder cells are better for acid solutions is, they are not so readily attacked by the acids themselves as the softer baked, and therefore last longer.

In the case of the Le Clanché and Sulphur Sal-ammoniac batteries, the chief thing to be guarded against is the crystallizing out of the secondary salts before referred to, within the pores of the cells, closing them up and sometimes cracking the cell. If a hard-baked porous cell is used for either of these batteries, its pores will often be closed right up, and what is for practical purposes an infinite resistance will be introduced into the circuit. It must be remembered that the porous clay itself has a very high resistance; so that the conducting path, for practical purposes, is by way of the liquid *in* the pores; and if these are closed by a solid body, such as a crystalline salt, there is no appreciable passage for the current at all.

The zinc used in all galvanic batteries must be the purest obtainable; and it is also wise to cover it with a film of mercury, by a process known as amalgamating, because an amalgamated zinc lasts usually much longer without attention than one not amalgamated.

On no account must any foreign metal, other than mercury, be present; since it forms, with the zinc and the liquid in which the plate is immersed, a small galvanic battery whose circuit is always closed, as with the oxide already referred to, but of far more importance, and will immediately proceed to eat a hole in the zinc.

Cast zinc, unless cast from the purest refined zinc, and with great care to avoid the entrance of such impurities, always contains these foreign bodies. Further, as is well known, it is exceedingly difficult to cast zinc perfectly homogeneous, and without different physical formations in different parts. Thus, if one portion be harder than another, galvanic action will take place between them. If there be any protuberances on the casting, they will give rise to galvanic action, between them and the main body of the zinc.

The amalgamation of the zincs is of far more importance in acid batteries than in the Le Clanché and Sulphur Sal-ammoniac batteries, as the action of the acid upon the zinc is stronger. In fact, though the process of amalgamation, properly carried out, would appear to take the mercury right into the substance of the metal, the acid follows it; and, after doing more or less work, according to circumstances, the zinc loses that bright appearance which the mercury gave it, and will behave just as if it had never been amalgamated, except that it will have become very brittle. In acid batteries, therefore, which are used for telephone, telegraph, and signal work, the zincs are now usually cast in the form shown in Fig. 18, which was first introduced by Mr. J. S. Fuller. The truncated cone stands in mercury, within the porous cell, and keeps itself amalgamated by capillary attraction, the mercury continually creeping up its surface. This will not, however, protect it from what is known as local action, viz., the galvanic action generated by the presence of foreign bodies or different physical condition.

Fig. 18.

Zincs, as they are termed, for Le Clanché and the

Sulphur cells, are generally made from rod drawn specially for the purpose. They are cut to the required length, drilled or slotted at one end, a strong copper wire soldered in, the zinc amalgamated and the top blacked. Rods of different thickness and length are used for the different sized cells. By some makers the copper wires attached to the zincs are covered with tape; but though this partially protects them from the action of sal-ammoniac spilt over them, it is only a partial protection, as the covering soon rots, and the wire is exposed as before; moreover, the sal-ammoniac may creep under the covering and eat the wire in two unseen. The Author prefers to have a wire sufficiently stout not to be quickly eaten through; and he finds that there is rarely any trouble from this cause, as the men in charge of the batteries soon learn to look out for the green chloride of copper which is formed, and to scrape it off, to wipe their connections over

Fig. 19.

after doing a battery up, and to be careful not to spill the acid or sal-ammoniac liquor.

The connections to the collecting plates are of considerable importance, and in some cases of some difficulty.

In the case of the copper plates used in Daniell's battery, there is not much difficulty. In the trough form of Daniell's battery, the copper and zinc plates are made about the same size, the former having a tongue

left on of sufficient length to cast into the zinc, as shown in Fig. 20, and to leave a strap to hang over the division between the cells.

Fig. 20.

In other forms of Daniell's battery, a strap is left on the copper plate and either clamped to the zinc or a hole punched in the copper strap to receive a terminal screw in the head of the zinc.

With platinum plates, as the metal itself is too expensive to allow of much in the way of straps, the platinum plate is usually clamped to the zinc, the latter being bent for the purpose.

Connections to carbon plates are the most difficult, carbon itself being brittle and not easily worked, and being also porous, so that unless precautions are taken to prevent it, the liquid in which the carbon is immersed, or some of the salts which are formed in the battery, will creep up and destroy the connection at the top.

Fig. 21.

There are two methods employed to obtain a connection to the carbon plate of a galvanic cell; one is, to cast a lead cap on the top of the carbon and a brass terminal screw in the lead cap for the attachment of the connecting wire (Fig. 21). The other is, to provide a brass clamp (Fig. 21*a*). In the early days of the Le Clanché battery, considerable trouble was experienced with cells breaking down for no apparent

reason. The cause was traced eventually to a formation
of carbonate of lead between the carbon plate and its
lead cap, owing to the liquid having crept up the carbon, as before explained, assisted by the endosmosic
action of the current itself—that peculiar phenomenon which carries the liquid in the direction in which it is going. Arrived at the
junction of the carbon with the lead, two
electro-chemical actions evidently took
place : first, there was the lead-carbon
battery with the liquid between, forming, as with the zinc and its oxide, a
battery with a closed circuit; but there
was also the action of the current itself
passing from carbon to lead, whenever
the battery was at work. It is hardly
necessary to follow the reactions which
took place. The result was, that a white
substance, having a very high resistance
in proportion to the rest of the circuit
and stated by chemists to be carbonate
of lead, gradually spread itself over the
entire connecting surface of the carbon-
lead joint, and for practical purposes
broke the circuit; that is to say, the
introduction of this highly resisting

Fig. 21*a*.

powder, directly in the path of the current, reduced
its strength so much,— in accordance with Ohm's law,
—that the apparatus ceased to work when connection
was made.

The difficulty has now been overcome by the simple plan of filling the pores of the top of the carbon, where the connection is made, with paraffin wax. Where that has been done, according to the Author's experience, the trouble has completely disappeared. It is an open question whether lead caps with brass terminals cast in, or removable brass clamps are best. For many years, the Author invariably used the latter, as they have the advantage that they can be easily removed and cleaned, and that they will answer for successive cells. The objections to them are, that they are more expensive than the lead caps; that they require a stouter carbon plate than the former—otherwise a workman is apt to break the plate in screwing the clamp on—increasing the cost and decreasing the space available for the oxidizing agent in a cell of a given size; and further, that in some cases, though they could be got at for cleaning quite easily, they were apt to require more cleaning than the brass terminals on top of the lead caps. Owing to this combination of circumstances, the brass clamp has fallen out of use for the present, except where large carbon plates are used, such as those in batteries for electric lighting.

The carbon plates themselves have also undergone great changes, under the pressure of competition. Years ago, only plates cut from gas-retort carbon were used, and they were undoubtedly superior to the manufactured carbons; but the cost of cutting the carbon from the irregular mass in which it is deposited in the gas retorts, the price demanded by the gas companies for a material they were in those days only too glad to get rid of, combined with the improvement in the manufacture of the rival plates, have entirely changed the conditions, and one now rarely meets with a gas carbon plate in practical work.

One great fault that the manufactured carbons had, which has since been overcome by the best makers, was, the hard scale which was formed on the surface of the plate in baking, which offered a very high resistance to the passage of the current.

Though carbon is acted on less by the exciting and oxidizing liquids, and the salts that are formed in the working of the battery, than any other substance except platinum, it is a mistake to suppose that it escapes entirely.

Being porous, for instance, the secondary salts, before referred to, often crystallize out within its pores; and, as seen in the case of the carbon and lead, some action may take place and in fact usually does, as an old carbon plate can rarely be used for a cell a second time, if it has done much work in the first. Moreover, carbon, like every other known body, is soluble by the current.

Chromic acid, also, and some mixtures containing nitric acid, act very powerfully upon carbon plates; so much so, that a cell in which chromic acid is used largely will not maintain a constant current, owing to the fact that the electrical resistance of the carbon plate rises very much from the action of the chromic acid upon it.

The batteries in practical use for telephone, telegraph and signal work, are the Daniell's, now almost
obsolete, the Mercury-bichromate, the Le Clanché, and
the Author's Sulphur Sal-ammoniac.

The Daniell is made in two forms, the trough (shown
in Fig. 22), which is still much used for railway signals
and telegraphs ; and the jar form, as follows :—

An outer containing jar, a porous cell, a copper

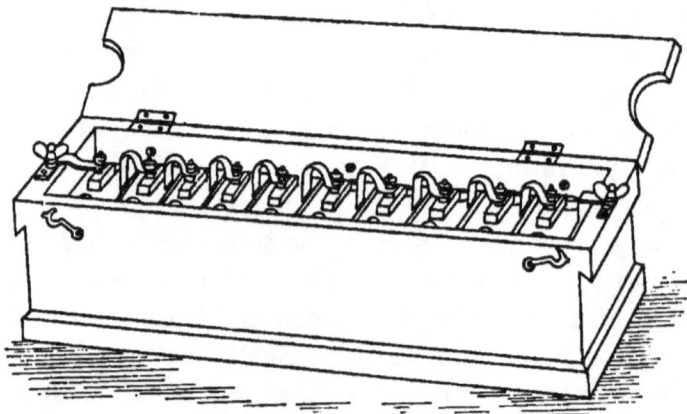

Fig. 22. Trough, Daniell's battery.

cylinder, either surrounding the porous cell or inside
it, and having a strap on a convenient part, as already
described, to connect to the next cell, and an amalgamated zinc cylinder which takes the place left vacant
by the copper cylinder.   If the copper is inside the
porous, the zinc is outside, and *vice versâ*.   With the
copper cylinder is a solution of sulphate of copper,
spare crystals being placed in any convenient manner,

so that they can drop down into the cell as required and maintain the solution at proper strength, during the working of the battery.

The zinc cell is usually charged with water, or a little weak sulphuric acid, and allowed to form itself into sulphate of zinc. The battery does not reach its full strength until a large percentage of zinc sulphate is formed. Battery men usually take some of the zinc liquor from old cells, to get up a newly-charged battery, but they do not use sulphuric acid, as the battery does not behave so well if they do, the action of the acid upon the zinc being too energetic.

Where the Daniell is used for large currents, of course sulphate of zinc would be used for the exciting liquid, or even dilute sulphuric acid.

The Daniell's battery, though it was justly hailed as an enormous advance on previous batteries, inasmuch as it enabled a continuous current to be maintained for the first time, has long since been superseded for most of the work for which it was formerly used, and for the following reasons :—

Local action, as it is termed, that is to say, the inter-action that goes on within the cell, irrespective of the current-passing, is always very strong in Daniell's battery.

As already explained, the porous division only separates the two liquids at first; gradually the sulphate of copper finds its way into the zinc portion of the cell, and the zinc is immediately attacked by it; zinc, as is well known, being always able to displace

G

copper from any solution in which it may be present. So we have a deposit of copper upon the zinc plate, forming once more the local cell with its circuit always closed, and its attendant troubles.

Copper is also often found within the pores of the porous cell itself; holes are eaten in the copper plate, and a greenish-grey compound, partly composed of sulphate of zinc and partly of sulphate of copper, works up all over the connecting straps, the sides of the cell, and forms other local circuits.

The net result of all this is, that the battery requires an enormous amount of attention compared to other forms, and it has therefore been gradually pushed out by cleaner and less troublesome cells. Added to the above, are the facts that the cell has an E.M.F. of only one Volt, where the other forms in use have one and a half, and that it is more expensive to work, irrespective of attendance.

The Mercury-bichromate cell consists of an earthenware or glass jar, more frequently the former, · containing within it a porous cell.

Inside the porous cell, standing in mercury, a small quantity being poured into the cell for the purpose, is the zinc, of the form shown in Fig. 18, page 74, having a stout copper wire, about No. 12 gauge cast in the top of the zinc and an eye formed in the other end of the copper wire, to slip over the terminal screw of the carbon plate of the next cell.

In the outer jar stands the lead-capped carbon already described, resting usually against the side of the jar and supported partly by the zinc wire of the next cell. In the same portion of the cell is the bi-chromate mixture, which consists usually of bichromate of potash with a certain proportion of sulphuric acid and water. The inner cell is generally charged with plain water, the acid and bichromate from the outer cell percolating through to the zinc as mentioned. The cell, when first charged, furnishes very little current, its internal resistance being so high, in consequence of the plain water in the inner cell. As this becomes acidulated by the percolation from the outer solution, the resistance of the whole cell gradually falls, and the proper current is furnished.

Some makers of bichromate batteries send out two kinds of powdered chemicals, a red powder for the outer solution and a grey powder for the inner one; the former is merely some special bichromate compound and the latter a salt of zinc.

When first charged, the solution in the outer cell of a Mercury-bichromate cell is of a deep orange colour; and when the battery is furnishing its proper current, that in the inner cell is the same, but apparently a thinner and weaker solution. As the cell continues to work, the solution in the inner cell becomes of a deeper and deeper colour; then the colour in both becomes a lighter orange, then green, and finally blue, when the solution is quite exhausted.

So long as the solution is of an orange colour, the cell will furnish a strong current for the work it has to do; provided the other parts of the cell are in order.

When it becomes green, the current is much weakened, both on account of the increased internal resistance of the cell, and of the decreased E.M.F., the secondary salts before referred to having begun to form, and to modify the resultant E.M.F. of the cell.

When the solution is blue, it may be taken that all or nearly all the oxidizing material is used up, and that the liquid in which the zinc stands is heavily charged with these secondary salts, such as chrome-alum. If the action of the cell be allowed to go on, dark blue crystals of chrome-alum will be found closely adhering to the carbon plate and zinc rod, and to the porous cell.

The cell is charged only with plain water, instead of acid, or with a sulphate or chloride of zinc, in order to reduce the action upon the surface of the zinc, and so preserve the clean amalgamated surface. If the action upon the zinc be allowed to be too energetic, the capillary action of the mercury will not be able to keep it amalgamated, and oxide will form, often also followed by a yellow deposit of some of the bichromate salts, with their attendant troubles.

The zinc rod in a bichromate cell, or, in fact, in any cell using acid, should be clean and bright, otherwise a deposit forms upon the zinc, and the strength of the current falls.

A zinc which pits, or forms irregular circuits in itself, owing to bad casting or impure metal, should be replaced, as it will be sure to give trouble.

The current furnished by a bichromate battery will occasionally fall, owing to a want of acid in the solution. It has already been explained, that plain water has about ten times the resistance for the same dimensions, as sulphuric acid; and the resistance of any solution in which acid is employed may be varied within very large limits by altering the proportion of the acid present. Thus, should there have been present any impurity with the bichromate, which has combined with a portion of the acid and withdrawn it from solution, the resistance of the cell would rise, and the current passing out would be weaker.

The only method of testing this, unfortunately, in the absence of an acidometer, is by the tongue. Every battery man knows when he has sufficient acid in his battery by the taste; but it is by no means an agreeable process.

The porous cell also, as mentioned above, may become hard; that is, its pores may be closed up, and will add a very large resistance to the circuit. Another troublesome fault in the porous cells of bichromate batteries is, the bichromate salts creep up the porous cell and form a crust outside the liquid at the top, sometimes even forming a circuit of high resistance between the zinc and carbon; but it being one that is always closed, forms a serious drain upon the cell. This difficulty is overcome by filling the pores of the cell with paraffin wax for about half an inch from the top.

The same thing may be done with the bottom of the porous cell, to prevent the zinc rod from being stuck to it by the crystals which form.

Chromic acid has been used in these cells in place of bichromate of potash and sulphuric acid; but it is doubtful if much is gained, as it so quickly attacks the carbon plates.

The mercury-bichromate cell is not so much employed as the Le Clanché, though many people prefer them, and a great number are in use. The reasons are, first, the cell is not so simple as the Le Clanché in maintenance. There is not the ease and certainty of finding your fault, and you need to have so many things handy.

Secondly, there is more wasteful action going on, or, as it is termed, local action; chemical reactions that do no good, and that waste material.

Thirdly, perhaps the greatest objection of all is the use of sulphuric acid. It is well known that sulphuric acid will attack all fabrics except those made wholly of wool, and will burn holes in them, leaving a nasty red stain till the hole is complete. Further, it will burn wood, cork, or the human flesh; all the above being due to its remarkable power of absorbing water, so that it is not nice to handle, especially where men may have cuts or abrasions on their hands. The battery is quite out of place for domestic electric bells, for instance, owing to this failing.

The Le Clanché battery is the one that has done most to forward the introduction of electric bells, mine signals, and other apparatus where the battery is or may be in unskilled hands, and where the minimum of attention is before all things the great desideratum.

It has, like the others, an outer jar of glass or earthenware. The favours are fairly divided in this case, each make having its advantage. The glass jars are made square, with a place for the zinc rod, as shown in Fig. 23. They are lighter, and they have this great advantage, that a battery man who will take the

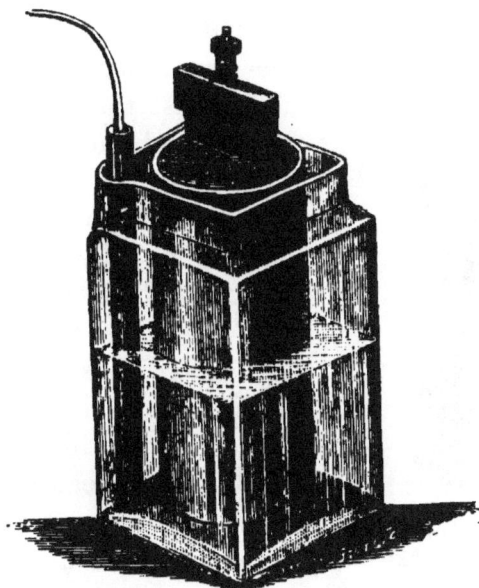

Fig. 23.—Le Chanché Cell, with glass jar and lead-capped carbon.

trouble can economize very considerably in sal-ammoniac, as, if his batteries are in the light, the cloudy state of his liquid will show him when to add more sal-ammoniac or to change his liquor.

On the other hand, the corners of the glass jars are very apt to get broken, so that a battery not unfrequently fails, owing to the exciting liquid having all run out of one jar. Moreover, they rather easily get broken in carriage and in handling, and they lend themselves very readily to the creeping action that will be described presently.

Earthenware jars are heavier and take up more room, but they do not break so easily ; and cracked jars are not often found, after the cells are at work, if the jars be of the best makers. There are now several makers of battery jars, who make them from a good vitreous clay, and put a glaze on them which will resist all acids. Figs. 23 and 24 show Le Clanché cells complete, with glass and earthenware jars.

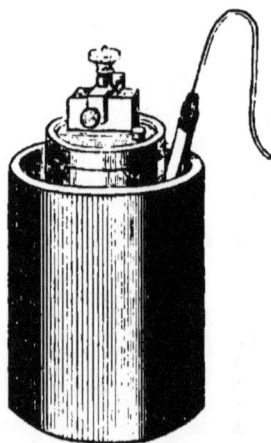

Fig. 24.—Le Clanché Cell, with earthenware jar and clamped carbon.

Standing in the outer containing jar is the filled porous cell, containing the carbon plate and the oxide of manganese that is to neutralize the hydrogen. The capped or clamped carbon plate is held centrally in the cell, and the space filled round with a mixture of granulated carbon and manganese dioxide. The former is the gas retort carbon crushed to the size of a pea or small bean, and the latter is the manganese ore, the richest obtainable, which is crushed to the same size. The mouth of the filled cell is closed with pitch, a glass tube being left in the top, projecting down into the mixture, to allow of the escape of the ammonia gas which is formed, and of the hydrogen gas, when it has not time to be neutralized.

The proper proportion of manganese to carbon, and the quality of both substances, have a most important effect upon the working of the cell. Carbon, it will be remembered, is a moderate conductor, oxide of manganese is a very much worse conductor than carbon, as all alloys or compounds are worse than the elements from which they are formed. Therefore, if a large proportion of manganese be used, the internal resistance of the cell is increased, though the staying power of the cell should also be increased, and *vice versâ*. Further, manganese of the best quality, that is, yielding upon analysis the highest percentage of pure oxide, is more expensive than either carbon or manganese ore that does not yield so high a percentage; consequently it is quite possible to make a cell of a given size that shall be apparently equal to another of the same size, and yet not do its work so well or for so long, by the simple process of using either inferior manganese or an undue proportion of carbon. It is obvious, of course, that such cells could be made and sold for less than those in which a proper proportion of high quality manganese was employed. The cell, having an undue proportion of carbon, would even test better when first charged, as its resistance would be less than that of the other.

In the outer jar is placed a certain quantity of sal-ammoniac, crushed either to fine powder or to about the size of a pea,—the latter preferably,—and the jar is filled to about one-third from the top with the purest water obtainable. Condensed steam water is always best, if to be had; failing that, soft water, not hard spring water, should be used; and the reason is the same as that given for purity in other components of batteries, viz., hard water contains other salts, such as lime and sometimes magnesium, in solution; and, therefore, if this water be used, these salts will take their part in the electro-chemical reactions that go on within the cell, giving rise to E.M.F.s and possibly adding resistance, so that the battery cannot be made to do its full work. Salts of lime are very apt to fill up the pores of the porous cell, and cause trouble in the manner already described.

For the same reason the sal-ammoniac used should be pure, and care should be taken that the machine used in crushing it does not introduce any impurity. Ordinary commercial sal-ammoniac often contains large percentages of the salts of other metals, which act very prejudicially upon the life and work of the battery. In particular, the Author has known of cases where the porous cells had repeatedly given way, scaling and breaking up after being in use only a very short time, and apparently for no reason; the cause having been traced eventually to either impure sal-ammoniac or to hard water, containing metallic salts in solution.

Standing in the sal-ammoniac solution is the wired amalgamated zinc rod which has been mentioned,

The full action of the battery is as follows:—Zinc combines with the chlorine gas and the oxygen gas in the sal-ammoniac solution, forming zinc oxide and zinc chloride, and ammonia gas and hydrogen gas are set free at the carbon plate. The hydrogen gas reduces the manganese dioxide to sesquioxide, forming with the oxygen, of which it deprives the manganese dioxide, pure water. The zinc chloride, zinc oxide, and ammonia gas form further combinations with each other, as zinc-ammonic chloride, oxy-ammonic chloride, and others, which are soluble in a saturated solution of sal-ammoniac.

When a battery is first charged, if the filled porous cell be placed dry in the sal-ammoniac liquor, no current will be furnished, or practically none, there being no connection between the liquid and the carbon plate. As the liquid percolates through the pores of the diaphragm, the resistance gradually falls, just as with the bichromate; but the reason is, that the solution of sal-ammoniac forms the conducting path not only to the carbon plate, but to every particle of the loose agglomeration surrounding it, the mixture itself offering a very high resistance indeed. To get a cell up quickly, battery men frequently pour some of the sal-ammoniac liquor into the porous cell, so as to wet the carbon and manganese mixture with it as much as possible; but even then the cell will not reach its full strength, or, to speak more correctly, the resistance of the cell will not fall to its proper limit, until the battery has been working some time and the electric current itself has carried the liquid to every particle. A good plan, if time permits, of getting a battery up, is to allow a weak current to pass through it for some hours by connecting its terminals to some resistance, such as the high resistance circuit of a detector galvanometer such as will presently be described.

The mistake must not be made, of pouring plain water into the inner cell, or " Inside," as it is termed; if it is required to get up quickly; for the same reason as in the mercury-bichromate and the Daniell's batteries when their zinc cells are only charged with water, they will offer such a very high resistance.

After the Le Clanché cell has been working for some little time, the liquid in the outer cell will have fallen, owing to the portion that has passed into the inner one, and may be filled up again, keeping it about two-thirds full. As the cell works, the zinc rod becomes coated with a thin powdery coating of some grey substance, probably the chloride of zinc, but apparently it does not seriously affect the working of the battery in this case, *provided the zinc is pure and well amalgamated;* but if any impurities be present, it helps to eat holes and to form all the troubles known as local action.

One caution should be given here, with reference to the quantity of sal-ammoniac to be placed in the cell. Some of the text-books give directions, that only a half-saturated solution should be used. It has already been pointed out more than once, that the secondary salts formed in the battery are soluble in a *saturated* solution, therefore it seems obvious that a saturated solution should be used, and that means should be taken to maintain it in a state of saturation, notwithstanding the demands made upon it by the working of the battery. The simplest method of accomplishing this is to leave some sal-ammoniac in the bottom of the cell, and to allow the liquor to take it up as it wants it, and as it will do if allowed; and this is what is usually done in practical work.

Where glass jars are used, in the light, a battery man may save some sal-ammoniac, as already explained, by removing the spent liquor when it becomes cloudy, and recharging.

The connection to the zinc rod is a very important feature in all batteries. If it is made too small, it is apt to break off, or get eaten in two; if too large, it adds unnecessarily to the expense of the cell, and is stiff and awkward to handle, wasting men's time. In the early days of the Le Clanché battery, galvanized iron wires were used, which invariably rusted in two at the point where the wire entered the zinc, or rusted nearly in two and then broke off. They were also very stiff to handle.

In the author's experience a No. 15 or No. 16 copper wire annealed answers best for all sizes of Le Clanché cells.

One of the greatest failings the Le Clanché battery has, is the formation of a crust of sal-ammoniac on the outside of the jar, reaching in time to the bottom. Whenever the battery is placed in a warm situation, the liquor apparently evaporates and then deposits on the cold surface of the jar, but leaves a fine capillary space between the jar and the white crust; and as, when the crust reaches the ground or bottom of the battery box, the outside and inside portions respectively form the long and short legs of a siphon, the result is that the cell gradually empties itself of its solution, which is poured out on the ground; and hence two serious evils arise. First, when any cell is empty of solution, it breaks the circuit just as if a wire had broken, so that the apparatus it actuates will not work, and its action gets weaker and weaker some time before this happens; and secondly, the liquid on the floor or bottom of the battery box, the liquid and crust on the outside of the jar, and the zinc and carbon standing in the cells on each side,—zinc in one cell and carbon in another, connected by the zinc wire above,—form a galvanic cell whose circuit, though of high resistance, is always closed. Therefore it is important that Le Clanché and other batteries should be placed in a cool situation; and if this, as sometimes happens, is not practicable, this white crust should be scraped off as fast as it forms, and the jars kept filled up. Partial relief is found in greasing the top of the jar; and Mr. Gent, of Leicester, has recently patented a jar of the form shown in Fig. 25, the channel on the top being filled with paraffin wax. The author is not aware how far this has been successful.

The great features in favour of the Le Clanché are, its extreme simplicity compared with that of other batteries, its cleanliness, the absence of any very troublesome material, such as sulphuric acid, the comparative absence of local action, the small quantity of materials used, and the small attendance necessary.

Its failing is, besides that already mentioned, it will not stand hard continuous work so well as other forms, notably the mercury-bichromate, though it will approach them to a very large extent if it be made large enough. It will easily be understood that the staying power of the Le Clanché depends upon two things, the neutralizing of the hydrogen gas and the solution of the secondary salts.

The evolution of the hydrogen and the formation of the primary salts will be strictly in proportion to the current passing; each Ampère, or fraction of an Ampère, delivering its equivalent of hydrogen, ammonia, zinc chloride, etc., each second.

Fig. 25.—Gent's arrangement of Le Clanché cell, preventing the formation of white crust.

Then it follows, that if the hydrogen cannot find oxygen at the same rate as it is itself evolved, it must be delivered free and set up its opposing E.M.F. Further, if the ammonia, zinc chloride, and zinc oxide cannot combine, and their resultants be dissolved as fast as they are formed, they

set up opposing E.M.F.s and resistance at the surface
of the zinc rod; the joint result being, that the battery
breaks down, fails to give a current, unless time be
allowed to recover itself; and it follows further, that if
the total quantity of current passing has been such as
to develop more hydrogen, ammonia, etc., than the
quantities of the other materials in the cell can dispose
of, the cell is worked out and requires renewing with
fresh sal-ammoniac and a fresh "inside" (filled porous
cell). It will be obvious that, within certain limits the
solution of the above difficulty is merely a matter of
size and quantity; given a certain current for a certain
time, and a certain quantity of manganese and of sal-
ammoniac can deal with it; further, the rate of
delivery of current that can be taken from the Le
Clanché is also a matter of time and quantity. Thus
a certain current delivers a certain quantity of hydrogen, etc., per second. A certain surface and body of
manganese and a certain quantity of sal ammoniac
liquor are necessary to deal with these bodies at the
same rate as they are delivered.

In practice this resolves itself into the axiom : Make
your cells as large as you can in proportion to the
current passing ; and as you do not often know what
that is, make the cells as large as you can. Thus the
Le Clanché cells are made in three sizes, pints, quarts,
and five pints or half gallons. In the early days of
the battery everybody used pints; and, except in
special cases, where the current was small and such
as the small cell could deal with, everybody had

trouble, and many were inclined to condemn the cell. Later, we got to the No. 2, or quart size, which answered very well for a great many purposes, but was costly where there was any leakage, the inside cells requiring to be replaced very frequently. Now, many of us use the largest size, because we find that it is cheaper for all kinds of work, even for that for which the small size would answer, because the larger ones go so much longer without attendance; and for all cases where leakage is present, the gain is enormous.

On the other hand, an eminent telegraph engineer— the superintendent of one of the large railways—told the Author he thought of adopting the small-sized cell; but he was designing beautifully delicate apparatus, that would work with very weak currents indeed, in order to enable him to do so; and further, he had his trained staff of battery men constantly looking after them.

The striking contrast in the small attendance required to the Le Clanché, as compared with others, will be gathered from the following. The Daniell's battery, under the best conditions, always requires looking at every month or six weeks; the mercury-bichromate from three to six months, the latter being exceptionally good work; and both of these batteries will require attention at their time and often before it, if they are doing no work. The Le Clanché will go, under unfavourable circumstances, from one to two years, and often longer, without even looking at, if the size of the cell be in proportion to the work it has to

H

do, except where it is placed in a hot situation. The
cells which work the bells in the author's house, which
get a great deal of work, as there are bells from everywhere to everywhere, have been in use nearly six
years, and have only had very little attention during
the whole period ; they are half-gallon cells.

The matter of the use of materials was also strikingly
illustrated by the Author's own experience in the early
days of mine signalling. As the signals were not much
in favour, he was obliged to contract at a low figure
to keep them going; and to save money and gain
experience he became his own battery man, with the
result that, although the batteries received the hardest
work that any battery was ever subject to, and he had
to go a moderately long railway journey to the collieries,
he succeeded in making the contracts pay, because the
material used was literally almost *nil*. All that was
required was care and attention.

A form of the Le Clanché battery, which has been a
good deal used, but is not in the writer's opinion so
good all round as the porous cell form which has been
described, is what is known as the agglomerate. The
mixture of oxide of manganese and carbon, instead of
being filled loosely into a porous cell, is crushed to a
fine powder, and formed into small blocks or briquettes,
under hydraulic pressure, some glutinous substance
being added to give cohesion. The carbon plate is
placed between two of these blocks, the three being
held together by two Indiarubber bands. Fig. 26
shows this form. It is evident that, since the porous

cell is dispensed with, the resistance of this arrangement must be less than that of the porous cell form; and in practice the results are very good at first, but it is found that the agglomerate form requires very much more attention than the porous cell form. The former will do harder work for a time; but whether worked hard or not, the blocks apparently soon become coated with a white crust of the secondary salts, and the resistance of the battery is thereby so much increased that it is necessary to clean this off periodically, and in fact somewhat frequently. The agglomerate cell will apparently do a lot of hard work before it requires renewing, provided this white crust be got rid of from time to time. In some cases, the zinc has been enclosed in a small porous cell, which of course robs the battery of one of its great advantages; but the general consensus of opinion of practical men is, that the porous cell form is best, because cheapest both in first cost and in attendance.

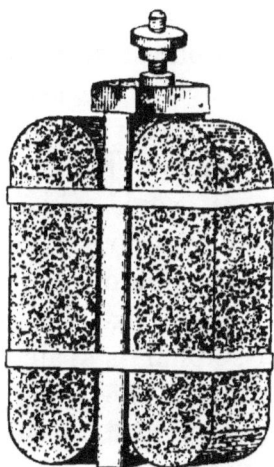

Fig. 26.—Agglomerate Le Clanché, inside.

THE SULPHUR SAL-AMMONIAC BATTERY

is similar in appearance and in nearly every detail to the Le Clanché. There are the same glass or earthenware jar, preferably the latter ; the same porous cell, with its clamped or capped carbon plate ; the same amalgamated zinc rod, with its copper wire soldered in the top, standing in a solution of sal-ammoniac ; but the contents of the porous cell surrounding the carbon plate consist of a mixture of granulated carbon as before, the place of the oxide of manganese being taken by granulated sulphur of about the same size as the manganese.

The purest sulphur only must be used, impurities in this having the same effect upon the working of the cell as impurities in the oxide of manganese or sal-ammoniac would on the working of the Le Clanché, viz., it would give rise to secondary reactions, that would create resistance and opposing E.M.F.s.

Less sulphur may be used than manganese in a cell of a given size, for the same work, since the whole of the sulphur is available for combination, while only a portion of the oxide of manganese is available, that portion which constitutes the difference in oxygen between the dioxide and the sesquioxide. It follows, therefore, that the internal resistance of the cell is less than that of the Le Clanché of the same size, owing to the larger proportion of carbon used, while its E.M.F. is practically the same ; that is to say, its E.M.F. when no current is passing. And as E.M.F. at the terminals of the cell is equal to the total E.M.F. created by the reactions of the cell (less the charge made upon it in accordance with the formula $E = C R$), it follows that the working E.M.F. of the sulphur-sal-ammoniac cell should be a trifle higher than that of the Le Clanché. Further, sulphur being cheaper than manganese ore, the cell costs rather less to make.

The action of the sulphur cell is as follows : The zinc combines with the chlorine of the sal-ammoniac, and with the oxygen of the water as before, ammonia gas and hydrogen gas being liberated within the porous cell. The hydrogen there forms with the sulphur, sulphuretted hydrogen, two atoms of hydrogen combining with one atom of sulphur ; but the action does not stop there. As soon as the cell is fairly up to its strength, zinc chloride is present within the porous cell, as well as the sal-ammoniac solution and the sulphuretted hydrogen. Hydrogen sulphide immediately forms with it zinc sulphide and hydrochloric acid, the latter again combining with the ammonia gas to form sal-ammoniac, the original solution.

The zinc sulphide apparently falls to the bottom of the cell as a white powder, and does not take any further part in the reactions.

One peculiarity of the sulphur cell is, that it takes longer to get up to its full working strength than the Le Clanché, this being no doubt due to the necessity for the presence of zinc chloride within the porous cell.

Another result, however, that usually accompanies this is, the cell appears to stand hard work better also. According to the author's experience, the cell will stand what is known as a dead short circuit, or, to put it more properly, a comparatively large current passing through it, longer than the Le Clanché. All that has been said of the Le Clanché, with reference to cleanliness, simplicity, convenience, and absence of local action or waste of material, when not in use, is equally true of this cell.

If a current of greater strength than the size of the cell warrants is allowed to pass for any time, somewhat similar results happen to those that occur with the Le Clanché. The hydrogen gas does not cease combining with the sulphur, no matter what current strength may pass ; but the hydrogen sulphide does not then enter so readily into its secondary combinations, forming instead within the porous cell, and after a time coming away freely. The presence of this hydrogen sulphide undoubtedly reduces the strength of the current passing, by introducing both resistance and an opposing E.M.F. ; but the E.M.F. generated by the hydrogen-sulphide-carbon galvanic couple in sal-ammoniac, is by no means so powerful as that of the hydrogen-carbon couple, and therefore the effect upon the strength of the current is not so serious. Moreover, unless the large current pass for a considerable time—that is to say, unless the supply of active sulphur is pretty well all used up, the battery is not usually in the exhausted state the Le Clanché is after a similar ordeal.

In other words, though the Sulphur Sal-ammoniac battery is not able, any more than the Le Clanché, to stand the passage of more than the fractional part of an Ampère for any of the regular sizes made, it will apparently stand a temporary accidental heavy drain for longer; and it moreover has the valuable property of giving warning, by the emission of the disagreeable odour attending hydrogen-sulphide, that of rotten eggs, when the proper working current is being exceeded. It should be noted, however, that for some work—such as where accidental short circuits are frequent, and the battery is placed in a living room—the sulphur cell would be inadmissible on account of the unpleasant smell.

Such cases should fortunately be rare; and the Author hopes, therefore, that his battery may have an extended use in the future. In his own work, which consists principally of mining signals—the hardest work, he believes, that any battery is subject to—it has, so far, answered admirably.

The whole of the above cells are used only for either very small currents or for apparatus using intermittent currents. They will not stand the continuous drain of a large current for many hours without breaking down. In fact, the apparatus should be arranged, if required to furnish a current continuously, for small fractions of an Ampère, $\frac{1}{10}$ or even less; the smaller the current the better the battery will stand up to its work. Larger currents may be, and are, used in practice, where the current only passes for a few seconds, such as in signalling on railways, in mines, with domestic bells, and the call-bells of telephone apparatus; because, though a large quantity of hydrogen and secondary salts are formed by the large current, in proportion to the time the current is passing, the interval of rest between the signals gives time for the hydrogen to be neutralized and for the secondary salts to be disposed of. In such work as the microphone circuit of telephone apparatus, where the current may be passing for several minutes together, the apparatus is arranged to work with as small a current as possible; though here, too, the interval of rest is more relied on than the weak current, as on very busy instruments, such as those employed in telephone exchanges, special batteries are always used—either very large Le Clanché or special forms of Mercury Bichromate or Daniell's; and they are kept up to their proper strength by frequent attendance.

Batteries naturally, therefore, divide themselves into two classes, those which furnish small currents, which have been already described, and those which furnish large currents, and which are used for electric lighting. The only forms of battery for large currents that need be mentioned are:—the Bichromate, a modification of the Mercury Bichromate, which has special solutions arranged to furnish a large current approximately constant for several hours; the Schanschieff, in which the mercurial salt already mentioned is used, and which has no porous cell; and the Lalande-Chaperon, in which caustic potash and oxide of copper are the solutions. As neither of these batteries is yet doing much practical work, outside the laboratory and special cases, such as miners' lamps, which will be more fully dealt with later on, it will hardly be necessary to refer further to them.

As already mentioned, a galvanic battery is merely a number of galvanic cells connected together. The E.M.F. obtainable with any galvanic combination is a fixed quantity: with the Daniell battery, 1 Volt; with the Le Clanché, Sulphur Sal-ammoniac and Mercury Bichromate, 1½ Volts. That is to say, the highest E.M.F.s that can be obtained between the terminals of any single cell of the types named, are the voltages given; and as we have before seen, the working E.M.F., that which actually exists between the terminals of the cell when a current is passing, and which is available for driving a current through that portion of the circuit external to the cell, is less than the figures named, by the charge made upon the E.M.F. which results from the combined electro-chemical actions, of the sum which results from the formula $E = CR$, E being the E.M.F. used in passing the current C through the resistance of the cell.

Now, these voltages, 1 Volt and 1½ Volts, would be quite useless for practical purposes, unless we employed very large conductors; and we therefore connect two or more cells together, and so add their forces, just as two horses attached to a carriage add their muscular energy, or should do so, to pull it along. To connect cells together so as to add their voltages or E.M.F.s, the + pole (carbon or copper), where the current leaves, of one cell, is connected to the − pole (zinc) of the next, and its own − pole to the + pole of the one behind it. The zinc wire is usually bent into a loop, which is slipped under the binding-screw on the top of the

carbon, and the latter screwed firmly down. The carbon pole +, at one end, and the zinc pole −, at the
other, are left free for connecting to the external circuit, and the usual plan is for two covered wires to be
brought down to the battery box, their ends bared,
scraped clean, one connected to the carbon pole by
being placed under the binding-screw and the latter
screwed down, the other firmly twisted round the end
zinc wire. Between the end zinc and carbon exists

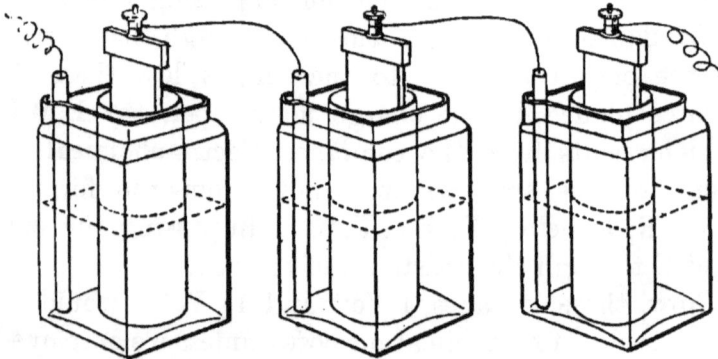

Fig. 27.—Le Clanché cells connected in series.

the full E.M.F. or difference of potential due to the
number of cells in combination (Fig. 27).

It is obvious, that by adding cells together in this
manner we may have as high a voltage as we please,
always remembering that we add the resistance of
each cell to the circuit, as well as its power of overcoming resistance, or its E.M.F. Thus, if we have 100
Daniell's cells, or 1000, we can have 100 or 1000 Volts

available to drive a current through the circuit formed by the cells themselves and the work outside.

Now, it will be obvious that the fact mentioned—that we add the resistance of each cell to the other resistance of the circuit, as well as adding its E.M.F.—may detract very much from the benefit that we receive, or it may not, according to the condition of the outer circuit. Take a circuit whose external resistance,—that outside of the battery,—is 95 Ohms, and a battery of 10 Le Clanché cells, whose resistance is 1 Ohm each. The current passing, by Ohm's law, will be $C = \frac{15}{95+10} = \frac{15}{105} = \frac{1}{7}$ Ampère.

Now add 10 more cells of the same size and resistance, and we get $C = \frac{30}{95+20} = \frac{30}{115} =$ about $\frac{1}{4}$ Ampère, and if our apparatus we require the current to operate be set to work with $\frac{1}{4}$ to $\frac{1}{5}$ Ampère, the addition of our 10 cells will have enabled us to do the work with 20 cells that we could not do with 10. But suppose the internal resistance to be 10 Ohms per cell,—as in the trough Daniell, and the voltage 1 Volt per cell, while the external resistance is only 10 Ohms.

With 10 cells we get $C = \frac{10}{10+100} = \frac{1}{11}$.

,, 20 ,, ,, $C = \frac{20}{10+200} = \frac{2}{21} = \frac{1}{10.5}$.

,, 30 ,, ,, $C = \frac{30}{10+300} = \frac{30}{310} = \frac{1}{10.3}$.

the current keeping practically the same, though we are employing 20 or 30 cells instead of 10, and using twice or three times the quantity of material. That is to say, we get only the same result at twice or three times the cost. If our bell is made to ring with $\frac{1}{11}$ Ampère, we do not get any practical advantage by using the larger number of cells, while, if it is made to ring with $\frac{1}{10}$ Ampère, it still refuses to do so. This is a case, of course, that would rarely or never occur in practice. It is introduced here in order to explode a fallacious argument that is given in nearly all the text-books, as to the proper method of connecting cells so as to get the best results.

No electrical engineer would arrange a circuit having all the resistance inside the generator, whether it be a galvanic battery or any other form. If he was unable to obtain any batteries except those having a high resistance, he would wind his apparatus with a gauge of wire that would reduce the current and at the same time give him the result he required with the smaller current; and with such a construction, that is to say, with internal and external resistance properly proportioned, every added cell would increase his available E.M.F. and working current. But on the above imaginary case, or somewhat similar ones, the text-books have built up a theoretical connection of cells, to get the best results, which is most misleading. It will be seen from the above, that if with our 1 Volt 10 Ohm cells, instead of adding their E.M.F.s and resistance, we join them in what is termed parallel or multiple arc,—that is to say, if, instead of adopting the series plan, shown in Fig. 27, page 106, we adopt that shown in Fig. 28, and connect the two batteries of ten cells with the two copper poles of the two tens together and the two zinc poles of the two tens together, and connect them to the external resistance, as shown in the figure,—it will be obvious that we have only the same voltage as with one set of batteries; but we have halved the resistance of the battery as a whole. Instead of $C = \frac{10}{10+100} = \frac{1}{11}$ Ampère, we have now with 20 cells $C = \frac{10}{10+50} = \frac{10}{60} = \frac{1}{6}$ Ampère, or we have nearly double the current; and if we use 30 cells in the same way, we have $C = \frac{10}{10+33\frac{1}{3}} = \frac{10}{43\frac{1}{3}}$, or about $\frac{1}{4}$ Ampère.

But although this plan looks well in theory, it will

Fig. 28.—Two trough Daniell's batteries connected in parallel circuit.

not work, and is never used in practice; and the reason is, that the batteries obey Ohm's law just as every other part of the circuit does, and that the current takes cognisance of *all* the conditions present.

It has been explained, that whenever an E.M.F. finds a path open to it, such that the current can pass through and come back to the starting-point, it drives a current through that path; and it has also been ex-

plained that either resistance or opposing E.M.F. will
prevent the passage or modify the strength of the current.

Now, in the present case we have, *in theory*, two
E.M.F.s equal and opposite, and two resistances exactly
equal; therefore no current *should* pass. In *practice*, we
have nothing of the kind; the E.M.F.s are *never* equal,
nor are the resistances; consequently, if the E.M.F.
at terminals of battery A be 10·5 Volts, while that of
battery B is 9·5 Volts, A will have a preponderance of
1 Volt, and will drive a current through both in the
direction of the excess E.M.F., the strength of which
will be $C = \frac{1}{200} = \cdot 005$ Ampère; a small current, but
always passing, and it must be remembered, in the
one-half of the battery passing in the opposite direction to the working current and assisting such actions
as the working of the sulphate of copper into the zinc
cell in the Daniells.

As a matter of fact, too, the difference in E.M.F. may
easily be more than the hypothetical case given; and
even if it be not at first, when the batteries are connected, the action of the current will soon make it so,
by the old method that has been so often mentioned,
of setting up opposing E.M.F.s and building up resistances. In making cells, no maker could ever possibly guarantee that any twenty cells, or even any two,
shall all have exactly the same E.M.F. and resistance;
though they may be made from the same material,
by the same hand, and that hand shall be as careful
as human hand can be. Nature does not make all the

grains of manganese ore the same; nor does the chemist make his sulphate of copper, or his sulphur, or his sal-ammoniac all exactly the same, grain for grain. Further, a slight extra squeeze of a lead cap in cooling, or in the pouring in of the mixture of carbon and manganese, may make sufficient difference to start this action. Cells which are perfectly alike for practical purposes, may easily have a difference of this kind, when pitted against each other, as in the arrangement of connections described.

It is for this reason, therefore, that the electrical engineer *never* connects his batteries in parallel circuit, as shown above, unless it be for a very temporary purpose, such as a few hours' lighting by primary batteries. If he cannot get his batteries of low resistance he makes his circuit outside of his batteries of high resistance. He makes his batteries of low resistance if he possibly can, except in special cases; but he does so by making them larger, having larger plates, more solution, etc., and he then obtains the double benefit of low resistance and lasting power in his cells; that is, the larger the cells are made, the larger current they will stand for a given time without attention, and the longer they will stand a given current.

The number of cells to be used for any given work is of great importance in electrical engineering, as it affects the working of the apparatus in two ways. If you have the cells too few or too small, your apparatus is constantly failing; but, on the other hand, every cell adds to the expense, both of installation and of maintenance, since the same current passes through each cell, consuming the same quantity of materials in each, or approximately so; and, further, each added cell increases the strength of the current passing through all the cells. If the cells be made larger too, the current strength is increased, owing to the reduction of the internal resistance of the battery, but the advantage gained by the larger cell usually more than counterbalances this. Perhaps there is no point upon which the practical man is more likely to condemn what may be called laboratory or book theory than this. He buys or makes an apparatus, say a trembling bell, that will ring with one cell, when connected directly to the battery. Perhaps he puts a hundred yards of wire in the circuit, and finds that it will ring then. He fixes the bell where it is only rung occasionally, and it apparently goes all right for some time; but if he places a similar bell in an office or an hotel where it is rung very frequently, and uses only one cell, though it rings very well at first, just as it did in his preliminary trials, by-and-by it fails; first, at times towards the end of a day's work, then more frequently. The customer to whom the bell was supplied, who was delighted with it at first, now thinks electric bells are very good—when they act. In many cases, the young contractor,—who possibly has only allowed for one cell in his contract,—does not attempt to repair the mischief at the source, he will prefer tinkering the bell a bit, cleaning the contacts, regulating the spring, cleaning the push spring contacts or the wire connections; and after a thorough overhaul of this kind, very probably the bell goes on ringing again all right, but not for long. In a few days or a few weeks, according to the amount of work the bell is called upon to do, the same thing happens; and the customer asks if this will be necessary every few days or weeks; if so, he is going back to the old bells, and so on. Possibly, the young contractor, in addition to having only one cell, has it of the smallest size made, and in that case his troubles are intensified. If he had used the largest size, the extra expense would have been very trifling, but it would have enabled his bell to hold out longer. If he had fixed two larger cells, he probably would have had no trouble at all for some time. Unfortunately, no rule can be given either for the number or the size of the cells to be used in each case, as these must depend entirely on the conditions of work, and must usually be a matter of individual experience. The failure of the battery was due to two causes, increased resistance and decreased E.M.F.; and it is exceedingly difficult to estimate the value of these two factors. As a rule, where the insulation is perfect, twice or three times the number of cells that theory would dictate should be used. Where the insulation is good and the work very hard, it may even be as many as eight or ten times. Much again depends upon the construction of the cell. If that be made with no reserve of power, a larger number will be required in the battery, and *vice versâ.* Two grand

I

rules should always be remembered, the larger the cells are, the longer they will go without attention ; and on the other hand, the more attention you can give to a battery, the smaller and weaker it may be. Further, you can make up your power either in the size of the cells or in their number, the former being usually the best plan.

Faults.—Perhaps the most difficult of all electrical work, and yet the most fascinating, is the discovery of Faults, as they are technically termed, or sources of failure.

The battery, where one is used, is the most fruitful in faults. It is the part the trained engineer flies to first.

To understand how to seek faults without loss of time, one had better understand what is meant by a fault. An apparatus ceases to work, or works badly. In most cases this will be due to the working current having fallen below that for which the apparatus is constructed. How has this happened, if the cause of the trouble is in the battery? As already explained, the current can be weakened either by loss of E.M.F. or by increase of resistance, or both, as one leads usually to the other. Thus, if the resistance of the battery be increased, the charge upon the E.M.F. for the current passing through it is greater, and therefore the E.M.F. available for overcoming the external resistance and driving a current through the apparatus, whatever it may be, which the current is to actuate, will be less.

The principal sources of increased resistance are *dirt* and the chemical reactions before referred to. Dirty connections always increase the resistance, for instance dirty wires under dirty screws. Another source of resistance is the evaporation of the liquid. Since the resistance of any body through which a current is passing varies inversely as the cross section, it follows that the evaporation of the liquid, by reducing this sectional area, will raise the resistance of the cell.

Sometimes also, an outer jar will be cracked, and the liquid gradually leak out, in which case the current gets weaker and weaker and finally ceases when the jar is empty, or nearly so, the resistance of that cell having increased practically to infinity.

When the battery is placed in warm places, too, the crust which forms and which has already been described, will empty a cell just as a crack in the outer jar would have done.

Other sources of resistance are the filling up of the pores of the porous cell, and of the carbon plate where carbon is used; the formation of highly-resisting crystals round the carbon, or round the carbon and manganese or sulphur; the formation of a crust upon the zinc; and lastly, of the reduction of the conducting power of the liquid, by the chemical actions that are constantly going on. Where sulphuric acid is used, as the acid is displaced the resistance of the liquid rises and the E.M.F. falls.

And now to find the fault. Personal inspection by an experienced electrician will often detect a fault

at sight ; but the more experienced he is, the less he
will be disposed to rely upon his eye, if he can possibly avoid it.   He will take the detector galvanometer
shown in Fig. 29 with him, and will usually find the
fault with unerring certainty in a few minutes.

The detector galvanometer, which will be described
more particularly under measuring instruments, is an
arrangement of insulated wire surrounding a vertically
suspended needle magnet, so that when the current
from a battery or other source of electricity is sent
· through the wire, the deflection of
the needle out of the vertical is
roughly proportional to the current
passing.   A piece of insulated wire
is connected to each of the two terminals, of what is termed the short
or quantity circuit of the detector,
the covering being removed from the
ends of the wires, the bared ends
scraped clean and firmly clamped
under the terminal screws.   The other ends of these
galvanometer wires, as they are termed, are bared and
scraped, and held, when testing, firmly between the
thumb and forefinger of each hand.   A place is scraped
clean with the knife on the zinc wire or strap of each
cell, the galvanometer placed where its dial can be seen,
and the engineer proceeds to test each cell in succession
by pressing· the bared ends of the galvanometer wires
firmly on the bared places on the cell connections, as
shown in Fig. 30,   A cell of a given size will produce

Fig. 29.—Detector
Galvanometer.

a certain deflection on a certain galvanometer while both are in order, so that by testing each cell in succession, in the manner described, those that are more or less faulty can be detected, and, what is more, the cause can be discovered. Suppose, as is usually arranged, that the galvanometer gives a deflection of 45° with a cell at its full strength. If the battery has

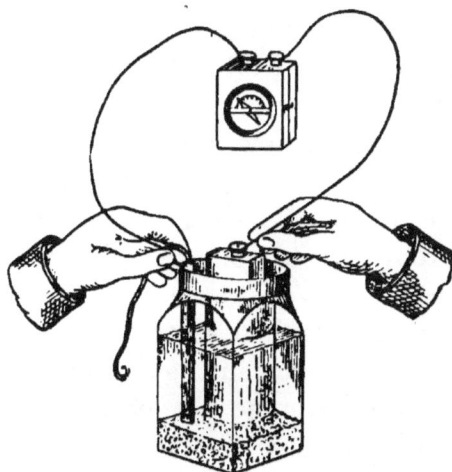

Fig. 30.—Testing a Cell with a Detector Galvanometer.

been at work some little time, probably some of the cells will now give 35°, 30°, 20°, some even 10°, and possibly one or two 0°. If the apparatus has ceased to work entirely, the replacement or removal of the cells giving 0° will generally put matters right, as they must be offering a very high resistance indeed; practically they are dead horses, stopping the others working by their inertia.

It may be, however, that the apparatus is only working weakly, and in that case several of the cells may
be giving 5° or 10°, while the rest give 30° and 35°;
or they may all be giving only 10° to 15°, all having
worked down together; in either case the removal of
the weak cells, or the renewal of their faulty parts, will
put matters right.

The next point is, to discover why any cell is weak.
Is it the zinc? the solution? the carbon? the copper
plate? the filled porous? etc. If the zinc is at fault,
it can usually be seen by simple inspection; or, to
make certain, place a new zinc in the cell, or scrape
the zinc well, and test again. If the deflection is
unchanged, it is not the zinc. Sometimes it is the
connection from cell to cell, the zinc wire or strap,
which has been allowed to become dirty. Remove
this, and clean the terminal screw, and test again. If
this makes no difference in the deflection of the galvanometer needle, examine the carbon itself, and, if
practicable, make a connection to the carbon plate below the cap or clamp. This failing, examine the solution. . With the Le Clanché or Sulphur Sal-ammoniac
batteries, if there is some sal-ammoniac on the bottom
of the outer jar, the solution is all right; if not, the
solution may be exhausted. Put some fresh sal-ammoniac in, cut the cell out of the circuit, if the deflection
is very low, and allow the saturated solution to work
into the inner cell, when the deflection will probably
rise, the failure being due to the secondary salts, which
have not been dissolved, crystallizing out within the

porous cell. In the case of the Mercury-bichromate, look at the colour of the solution; if bright orange, taste for acid; if that is right, the porous cell is probably hard.

With the Le Clanché and Sulphur Sal-ammoniac batteries, if the zinc, carbon, connections, and solution are right, the filled porous must be at fault. Change it, or let it rest. Sometimes a cell will get up with rest, especially if it has had a temporary short circuit. Frequently the porous cell becomes cracked in working, owing to the expansive force of the crystallizing salts ; and the conclusion is erroneously jumped at that the cracked cells are faulty ones. This is not usually the case ; the cracking of the porous jar often saving it, by lowering the resistance of the cell. If the pores were to fill up instead of bursting the cell, then the internal resistance would rise, as the area open for the passage of the liquid would be less.

In some cases, however, where impure materials are used, the porous cells may be cracked, and the cells down as well, owing to the chemical action of the foreign matter.

Primary galvanic batteries, that is to say, those batteries in which electrical energy is furnished by the direct consumption of zinc and other materials, have not, as yet, been much used for electric lighting, owing to the cost per lamp per hour, or per Watt per hour, being so very much higher than the cost when furnished by a dynamo, driven by a steam or gas engine; and also from the great difficulty experienced in maintaining the current at a uniform strength. With batteries which are to furnish electric light, the conditions to be worked to are quite different from those ruling with the batteries that have been already described as in use for telephones, telegraph signals, etc. With the latter, all that is required is that the apparatus shall work; it will not matter if a bell rings a trifle louder at one time than another, so long as it rings loud enough to be heard; but with electric lighting, the current strength must be absolutely uniform within certain limits, or the battery furnishing it is useless. No one would tolerate a glaring light during the early hours of the evening, and a very poor one later on. The difficulty is caused solely by the alterations in the resistance and E.M.F., due to the chemical reactions that have so often been referred to; but with this difference, the battery that is used for electric light always furnishes a large current, while that used for telephone and signal work uses a very small current, therefore any variation in the resistance of the battery will have far more effect with the former than with the latter, owing to the operation of Ohm's law.

The attendance also upon primary galvanic batteries used for electric lighting is a very serious item. All the difficulties that have been mentioned in connection with other batteries are present here, in a greatly intensified form; and it is usually necessary to attend to the solutions every few hours, instead of every few months or years.

Some inventors have resorted to the plan of introducing secondary batteries, connecting them directly with the primary battery, and the lamps with the secondary; and this is done on the plea of economy, the actual reason being, however, that, as will be seen later on, the secondary acts as a regulator and reservoir to the primary. It will receive the current from the primary day and night, and so apparently avoid the waste which always rules in a primary battery when it is not doing useful work, as a primary battery can never be at rest. It also maintains the current strength more uniform in the outer circuit, the lamps, etc. But the economy is not real. In fact, the introduction of the secondary battery causes more waste, though it undoubtedly covers some of the faults of the primary battery, if rather expensively.

There is one portion of the electric lighting field that it is to be hoped primary galvanic batteries may be useful in, viz., that of the Miner's Portable Electric Lamp. Though the cost of zinc and acids used in producing the light here bears the same or even a higher proportion to that of coal used in driving a dynamo, as in other cases; yet, as the quantity itself is small, and as, moreover, it has the advantage of being made into a portable form, whereas the engine and dynamo cannot, it is hoped that in this field it may do good work.

The *sine quâ non* for a miner's lamp is, of course, that the battery shall furnish a fairly constant current for the time of a miner's shift, including the time occupied in going to his work and returning from it.

CHAPTER V.

ELECTRIC BELLS are of two kinds, those known as trembling or vibrating bells, which ring continuously, as long as a current of the required strength is passing through them, and single-stroke bells, which ring once each time the circuit is completed.

Trembler bells are used principally for domestic purposes, and occasionally for mines or fire alarms. Single-stroke bells are used more for mines and railway signals, where a clear, distinct signal is of importance.

The construction of each is nearly the same, the differences being in the form of the apparatus, and in the arrangement added to the trembler bell to cause the hammer to vibrate.

In each there is an electro-magnet made in four parts, two limbs of round iron carrying the coils, a flat back or yoke piece, sometimes made part of the frame to which the electro-magnet is fixed, and the movable armature with bell-hammer attached.

In the single-stroke bell, as shown in Fig. 31, the armature is pivoted, and either it or its hammer shaft works between regulating screws, so that its distance from the poles can be adjusted. In some forms, as where the bell-dome is carried under the electro-magnet, the pull of the latter is balanced by a spring whose tension can be regulated, the object being to pull the armature back after striking. In those forms in which the bell-dome is over the electro-magnet, a straight steel spring is usually provided for the same purpose;

Fig. 31.—Inside of Single-stroke Bell, showing Electro-magnet and Hammer and Regulating Screws.

but it does not come into action until the blow has been struck. This form is decidedly the best.

In the trembler bell the armature is attached to a short, broad steel or brass spring, of sufficient strength to pull it away from the magnet poles when the current is not passing. The other end of this spring usually forms a current-breaker, or contact piece. When the bell is not ringing it rests against an adjustable stop, placed for the purpose, as shown in Fig.

32. When the current passes round the electro-magnet, the armature, being attracted, causes the hammer to strike the bell, the latter being placed so that it can do so; but it at the same time causes the contact spring before mentioned to leave its contact point, and thereby break the circuit. The circuit being broken, the attraction of the armature ceases, it falls back from the poles, remakes contact, and is again attracted; the result being the continuous vibrating ring, similar to the old clock alarum, with which every one is familiar.

It will easily be understood that the electro-magnets and other parts of electric bells should be properly proportioned to the work they have to do. A large, deep-toned mining or fire bell,—whose ring is required to be heard either at a long distance or above the noise of moving machinery,—requires a harder blow than a small bell, such as is suitable for the kitchen or servant's bedroom of a private house; and, again, the hall of a large hotel may require a bell not as loud as a mining bell, and yet louder than one that would answer for a private house.

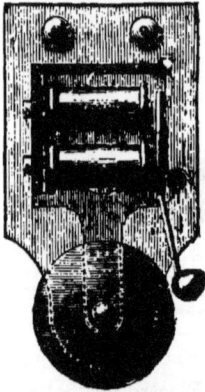

Fig. 32.—Trembler Bell, showing vibrating Armature, Hammer, and Contact Screw.

A little consideration will show that, in order to provide the requisite strength of blow, two things principally have to be considered—the size of the electro-magnot itself, and the exciting power; and in practice it works out cheaper to make your electro-magnet large and powerful, than to have more power elsewhere. It has already been fully explained, that the power you get out of a given electro-magnet depends directly on the exciting power and inversely on the magnetic resistance offered to that power by the magnetic circuit, consisting of the two legs, back, and armature; and it has also been explained, that with a given sized core of iron, the useful quantity of wire, and therefore the exciting power you can apply, is limited by the fact that each succeeding layer is larger and larger, offering greater proportional electrical resistance; while, being farther from the iron, its magnetizing effect is less. Therefore, if you make your electro-magnet small in proportion to the work it has to do, you can only get the necessary strength of blow on your bell by increased current strength, meaning an increased number of cells, increased cost of maintenance, and more frequent attention.

Perhaps there are few points in connection with electrical engineering so little appreciated by outsiders as this. It is so difficult for them to see how they can possibly be saving money by paying twice or three times the sum for an electric bell that they need do; and yet, if they were to take two bells, one with the electro-magnet large and powerful, and the other with its electro-magnet small and weak, and carefully test them, checking the total working cost in each case, the result would be somewhat surprising. The weaker the magnet, the greater must be the current strength; and so, where either magnet or battery are weak in proportion to the work to be done, the whole apparatus is an endless source of trouble. The battery is continually failing, do what you will, and continually requiring attention and renewals.

Another necessary feature of the magnets of electric bells is, that the cross section of the iron employed should be as large as possible; thin cores entail more

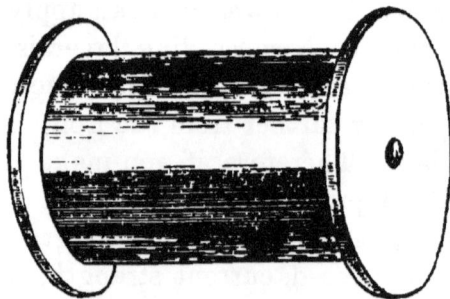

Fig. 33.—Forged Iron Bobbin for Electric Bells.

work upon the current, and therefore upon the battery, than thick ones. Further, the pole pieces and the armature should present as large a surface as possible to each other, though the former must not thereby be brought too close together, as the lines of force—which obey the law of magnetic resistance here as elsewhere, —will find a leakage path that will appreciably reduce the strength of the field between the poles and the armature, and therefore the pull upon the latter.

In the electro-magnet used by the Author's firm, and

worked out by him, the iron core of the coil is made to form the end or collar which holds the wire in and the pole piece as well, the bobbin or complete magnet leg being formed either by taking a piece of round iron of the full diameter that the wound coil will have, and turning it down between the ends, as shown in Fig. 33, leaving a collar at each end, or by forging a piece of iron to the same shape.

There are several advantages in this form; but the principal is —the magnetic resistance of the air space between poles and armature, and that of the joint at the back, are so much reduced, owing to the increased sectional area. At the same time there need be no fear of leakage from pole to pole, as these can be placed a considerable distance apart, without materially affecting the gain. It will be noticed also, that the exciting wire of the

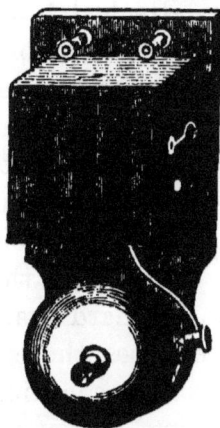

Fig. 34.—Trembler Bell, with Wood Cover on.

electro-magnet comes right up to the pole piece, so that the full possible work is got out of it.

Perhaps a more important point than even the dimensions of the electro-magnet is, that proper provision should be made that changes of temperature or other causes do not put it out of working order. Trembler bells are usually mounted on wood bases, arranged to hang vertically, with a cover over the electro-magnet, the hammer shaft protruding through a hole in the cover provided for it, as shown in Fig. 34. In the early days of electric bells, the electro-magnet and its armature, with their brackets, were mounted separately on the wood base, the most common type being that where the back piece of the electro-magnet was attached to a brass angle bracket, the latter being screwed to the wood base. The spring carrying the armature was also attached to a brass angle bracket screwed to the base, and a second bracket of similar form carried a second smaller steel spring, which, when the bell was not in use, pressed against the back of the armature, forming the contact-breaker. In those days, there used to be frequent complaints that bells did not ring; and often, on going to ascertain the cause, it would be found that the armature, or one end of it, was jammed hard against the poles and could not move; in other cases, the contact spring had worked back, owing to the continual vibration, and was just *not* making contact, so that no current could pass. One minute's work was usually sufficient to put the matter right; but it was exceedingly annoying, and in some cases expensive, when it happened frequently, where a railway journey was necessary. The iron springs that were used in those days were also a possible source of failure, owing to their liability to rust. The whole of these difficulties have been overcome by : first, using brass or well-made steel instead of iron springs, and secondly by mounting the whole apparatus upon an iron framework, as shown in Fig. 32, page 124, the contact pillar being insulated from the iron base, though supported by it.

The matter has even been carried further, some forms of mining bells having a complete iron case, the terminals and contact pillars being insulated from the case by collars of vulcanite or vulcanized fibre. In America, too, they are even making small domestic bells on the same lines.

In using these iron-cased bells, it must be remembered, of course, that the iron base and iron cover are conductors, and therefore the bared ends of the wires used to connect the terminals to the wire on the electro-magnet, or to the outside wires, must not touch the iron case. If they do, no current will pass

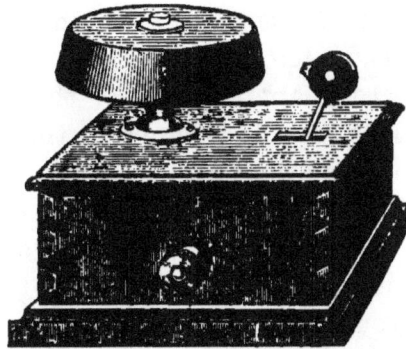

Fig. 35.—Single-stroke Mining Bell.

through the bell coils, all going by way of the iron case; the resistance of the latter being so small in proportion to the former.

With single-stroke bells precisely the same difficulties occurred in the early days as with tremblers. The electro-magnet and the supports of the armature were screwed direct to the wood base; the bell dome being carried by the cover, as shown in Fig. 35, and the hammer protruding through a hole made for it. After a

K

little time, possibly, the bell refused to ring, and its armature was found canted round, with one end touching one pole and the other end pointing away, so that it could not move; and in this case the remedy was not always so easy, if it was a large bell. The same arrangement has overcome the difficulties in this case as with the trembler bell, the whole of the working apparatus being mounted on a metal base, as shown in Figs. 31, etc., page 123.

A mistake that is sometimes made in the construction of electric bells is, the armature and hammer are made too heavy. This construction is based upon a wrong appreciation of the principles of the electromagnet. It is supposed that by having a greater mass in the armature you obtain a greater pull, because the magnetic resistance is thereby lessened. But though you undoubtedly lessen your magnetic resistance a little by, say, doubling the thickness of your armature, you nearly double the mechanical work that the current has to do; so that, setting off this heavy loss against the trifling gain on the other side, your advantage is a minus quantity, and you will find that your battery requires very much more attention in consequence; and unless it is made originally of very much larger *sized* cells than would answer with a lighter armature, it is constantly falling below working strength.

It must never be forgotten that it is your *battery* which suffers for all these mistakes or false economies, such as that of too small electro-magnets, though there may be nothing to show it, and you may fancy that you have got hold of bad batteries, whereas you have an improperly constructed bell.

The same remarks apply to the hammer. It is often imagined that you obtain a harder blow by using a heavy hammer; but practical experience shows that you do not, and the reason is a very simple one, viz., that in increasing the weight of your hammer, you increase the dead work which the current has to perform, more than you increase the blow. The blow would be harder if pulled up with the same force as the lighter hammer, but it is not; the product of the velocity with which the hammer approaches the bell, multiplied by its weight, which is the measure of the force of the blow struck, decreases after a certain weight is reached.

Domestic Electric Bells.—The earliest practical application of electric bells was naturally to domestic purposes. As soon as it was discovered that a bell of the form described might be made to ring at will, by connecting it with a battery and some form of circuit closer, as described in the chapter on the electric circuit, it was applied at once to communicate from dining, or drawing, or bedroom to kitchen; from principals' office to clerks, etc.

It will be seen that all that is required for this purpose is—that the bell shall be placed where it is required to ring to; the battery where convenient; and the circuit closer,—or push, as it is termed, from the fact that contact is made by pressing or pushing a

centre piece,—should be at the place it is required to
ring from; and that the three,—bell, battery, and push,
—should be connected by wires insulated from each
other, so that no contact can be made except that at
the push, in obedience to the will of the operator. In
practice this is accomplished by placing the battery
in some out-of-the-way spot, leading two covered wires
away from it, connecting one of them to the bell, and
taking a wire from the other side of the bell and
leading it, with the one from the
battery, to the push.

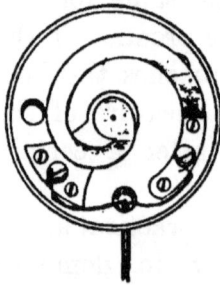

The usual form of push is shown
in Fig. 36; it consists of two German silver springs, attached to
a circular wood or vulcanite base
by screws. One spring remains
usually close to the wood base, the
other being arranged, by its form,
to stand away from it, until brought
into contact by pressure of the
centre piece. A wood, china, or
metal top, with a hole in the centre, screws on to the
wood base; and through this hole in the centre projects the thumb-bit or centre piece, working freely, so
that on the pressure of the finger being applied, it
brings the two German silver springs into contact, and
thus closes the circuit.

It has been mentioned in the chapter on the electric
circuit, that any number of circuits may be formed,
having the bell and battery in common and part of the

Fig. 36.—Inside of Push,
showing Springs and
Connecting Wires.

wires, but having separate circuit closers. This is taken advantage of to arrange communication between several rooms in a house and the kitchen, the wires being branched.

Each branch leads to its own push, and each push completes a circuit of its own from the room in which it is placed to the kitchen, and including the battery and bell. This is the arrangement used generally for small houses of about six or eight rooms, where the servant cannot make much mistake, and has not far to go to find out who rings.

The arrangement is rendered more certain by giving each room a number, and pressing the push in that room the number of times corresponding. Where necessary, also, the front door may have a separate bell of its own, another branch circuit being formed for it, in-cluding its own push or pull, its own bell and the battery, as shown in Fig. 37.

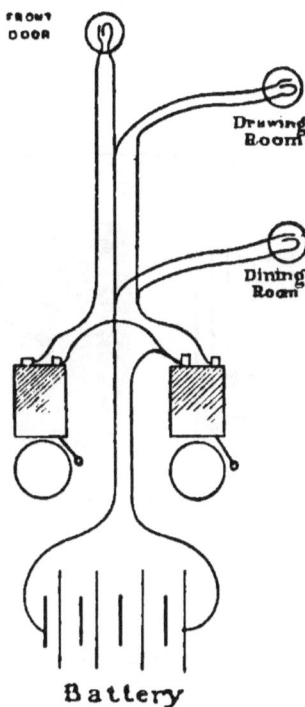

Fig. 37.—Diagram showing the Connections for Ringing two Bells separately from different Pushes.

For large houses, however, and for hotels, the above

simple arrangement would be unsuitable. A servant could hardly be trusted to be sure of the difference, say, between twelve and fifteen, even if ladies would have the patience to ring as many times. It is necessary, therefore, to make other arrangements to indicate the room calling. This may be done by giving each room a separate bell, as in the old days of

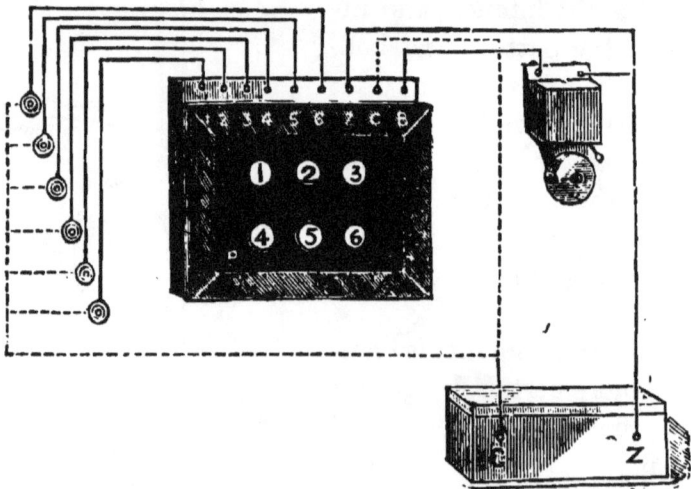

Fig. 38.—Indicator, with Bell, Battery, and Pushes.

the pull bell; but the plan does not work out very well. It is difficult to obtain sufficient variety of sound, and there is not the indication of the bell swinging, as with the old bells.

The plan usually adopted is, to give each room its own branch circuit, which shall include the bell and battery, its own push, and, in addition, an electro-magnetic arrangement that shall indicate which room has rung.

The whole, or a number of these electro-magnetic arrangements are placed together in one case, and behind a gilded glass, each one operating some form of tell-tale or indicator, and making it visible opposite

Fig. 39.—Thorpe's Semaphore Indicator.

a hole or space left in the glass. Fig. 38 shows an indicator with battery and pushes connected.

There are various patterns of indicator, some depending for their action upon the properties possessed by permanent steel magnets, others consisting of trip actions worked by the magnetism imparted to the

electro-magnet, attracting its armature and releasing
the tell-tale, which falls opposite its hole in the glass;
others, again, consisting of mechanical motions worked
by the moving armature, and bringing some disc or
shutter into sight. Others, again, are arranged upon
what is known as the pendulum pattern. A flag or
disc hangs opposite each hole, at rest, except when a
current passes through its electro-magnet. When this
happens, the flag or other tell-tale oscillates as a pendulum for a certain time. The object of the pendulum
indicator is to avoid the necessity for replacing the
tell-tale in its original position after its call has been
seen. It is obvious that with all the others this must
be done, otherwise there would be no indication when
the same room rang again.

The methods of replacement are also various, some
depend upon the before-mentioned property of the
permanent steel magnet, of being attracted or re-
pelled by unlike or like poles of an electro-magnet. A
current in one direction from the room causes the tell-
tale to appear; a current in the opposite direction, sent
by a push attached to the indicator case, causes it to
return to its original position.

In every other case, the tell-tale is replaced by a
mechanical arrangement, worked by a handle at the
side.

A favourite form, if well made, is that shown in
Fig. 40. The tell-tale is attached to a semicircular
steel magnet, which has either end attracted by the
electro-magnet, according as the current makes the

near end of the latter N, or S. It is doubtful whether the electrical replacement is best, as it necessitates rather a complication of connections inside the indicator, that may give trouble.

The reason this form works best when properly constructed is, probably, the small amount of work

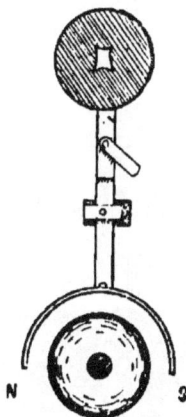

Fig. 40.—Indicator, with Half-hoop Magnet.

Fig. 41.—French Pattern Indicator.

the current has to do; though, as will be seen, the resistance of the magnetic circuit is rather high.

The form shown in Fig. 41, where a straight bar magnet works between the two poles of an electromagnet, is not seen much in this country, but was a great favourite in Paris. It did not work so well as the one with single electro-magnet, because the permanent magnet itself was somewhat heavy, and the tell-tale it carried was so also. Possibly, if both had been made lighter, and the electro-magnet poles nearer together, better results might have been attained.

In the form shown in Fig. 39, page 135, the armature of the electro-magnet engages with a ledge in the shutter and keeps the latter up out of sight until a current passes. On being attracted by its magnet, the

Fig. 42.—One form of Trip-action Indicator.

Fig. 43.—Paterson's Direct-action Indicator.

armature disengages itself from the shutter, and the latter falls opposite its hole in the glass. It is replaced by a pin upon a brass rod turned by a handle.

Fig. 42 shows another form of trip indicator. The great difficulty in connection with these trip indicators lies in the fitting. If they engage too tightly, the shutter will not be released when the armature is attracted. If they engage too lightly, they may fall when not wanted to, say from the vibration caused by passing traffic, a heavy person walking across the room, etc. ; further, they are apt to require looking at occasionally, to see that the catch acts properly, as it may wear from continually falling and replacing.

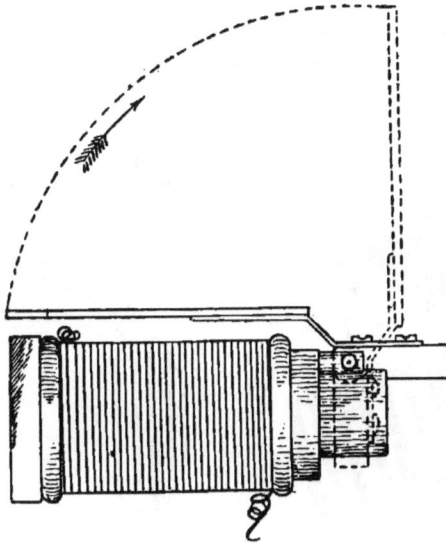

Fig. 44.—Side-view of Indicator shown in Fig. 43.

Figs. 43 and 44 show the only form the Author knows of in which the current is directly made to pull a shutter up. The electro-magnet has two limbs, and is fitted with semi-cylindrical pole-pieces as shown. The armature is pivoted just above the pole-pieces, and carries a light framework with a white paper on it. When at rest, the white paper which forms the tell-tale, lies back on the electro-magnet, overbalancing the armature, as shown in Fig. 44. On the current passing, the armature is pulled down by the attraction of the pole-pieces, overbalancing the tell-tale, and the latter comes up opposite its hole in the glass. This form absorbs more energy than any other, and it is usual to fit a small relay in the indicator case, so that the indicator can be worked directly from the battery and not weakened by the resistance of the bell, its trem-

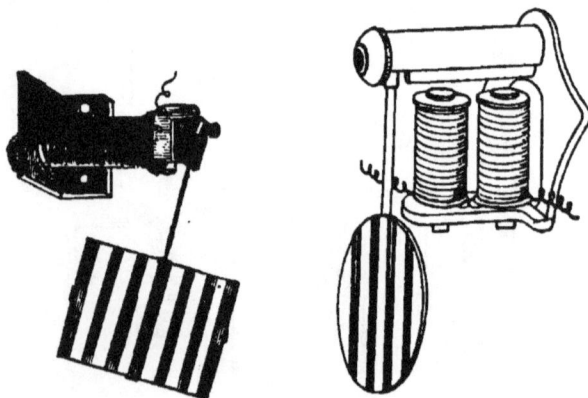

Figs. 45, 46.—Pendulum Indicators.

bling contact, and the wires leading to the push. Even then it requires more battery power, a larger number of cells, than other forms. The tell-tales in this case are replaced by stretched wires for each row, worked as usual by a handle at the side, and it has the advantage over most others, that the indicator may be placed higher up than is usual, the replacing arrangement being reached by a pendant cord.

Figs. 45 and 46 show the arrangement usually adopted for pendulum indicators.

The same observations that were made with regard to the construction of electric bells apply to indicators, except that the conditions are such as to preclude the employment of large electro-magnets, and therefore the work the current has to do must be reduced to the lowest possible limit. It is necessary with the indicator, as with the bell, that the electro-magnet and its armature and the tripping arrangement, shutter, or whatever tell-tale is used, should be all mounted together on a metallic base, otherwise trouble may ensue. In the early days of domestic electric bells, the unequal motion of different parts of the apparatus, owing to the wood warping under different temperatures, and locking the armature or the tripping arrangement, caused many failures.

It will be obvious that, on the principles enunciated in the chapter on the electric circuit, once we have a battery in the house, we may make as many branch circuits as we please passing through it, and each one may have not only its own push, but its own bell, which may be placed wherever we can carry an insulated wire, and that the indicators may be grouped as we please. Thus, in a large hotel it is frequently arranged that all the rooms on one floor ring to an indicator under the control of the chamber-maid on that floor. And even in private houses, it is frequently convenient to have a bell in the servants' bedroom to ring from the mistress's.

In one house that was fitted by the Author, there were three pushes in the owner's bedroom—to the kitchen, the servants' room, and the nursery. Any push may also be arranged to ring two bells at once, as,

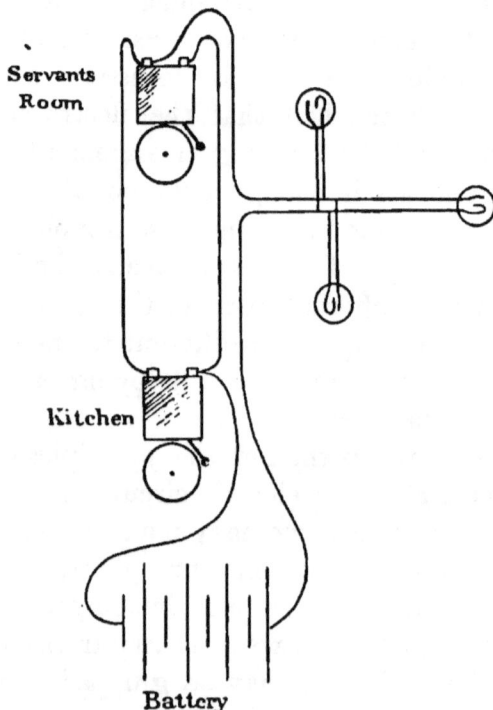

Fig. 47.—Diagram showing two Bells in Series, ringing from several Pushes.

say a bell in the kitchen and one upstairs, so that the servant can hear whether she is upstairs or down. In this case two branch circuits may be made for the two bells, each having outside of it, and in connection with it, the battery and push, as shown in Fig. 47. Or the circuit may be continuous through both bells, battery, and push, as in Fig. 48. The arrangement works best with the bells in separate branches. Also, where indicators are used, there may be as many pushes as you please in each room, branch circuits being taken to each push, so that each controls the circuit containing the bell, battery, and its own indicator.

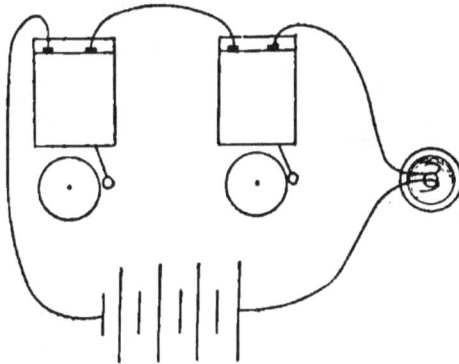

Fig. 48.—Two Bells ringing together, Connected in Series.

Another form of apparatus that is occasionally used in domestic work is the continuous-action bell, one arrangement of which is shown in Fig. 49. It is intended to go on ringing when once started by the current, until stopped by the hand; and is frequently placed in a servant's bedroom in such a position that she is obliged to get out of bed to stop it, and may then be supposed to have awakened sufficiently to get up.

The author has, however, more faith in a silk bell-rope, with a pear push at the end, lying on the pillow, that can be rung at will and for as long as desired.

In the continuous-action bell, the first action of the current, attracting the armature of the bell, releases a trip action something similar to some forms of indicator described; and the trip arrangement, on falling down, closes a circuit, including the battery and its own bell, so that the latter goes on ringing until the contact is broken. In some cases the contact is separate from the bell, and this, though more expensive, is also more reliable.

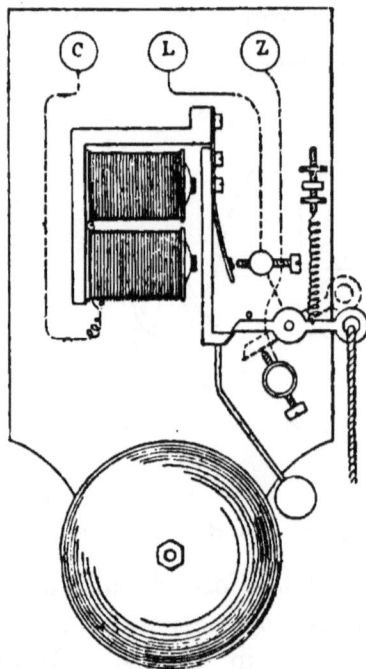

Fig. 49.—Continuous-action Trembler Bell.

The great objection to this form of bell is, that it is necessarily delicate in construction. If the catch is made very tight, the armature may not release it; and if it is made delicate, it is apt to be damaged by the rough hands of the servant maid, wakened out of her sleep, perhaps in the dark, and taken that this would be all the more likely if she had an inkling that it was within her power to prevent the bell ringing next morning.

Arrangement of Wires.—Where indicators, or annunciators as they are sometimes called, are used, or there are cross connections to various branch circuits,

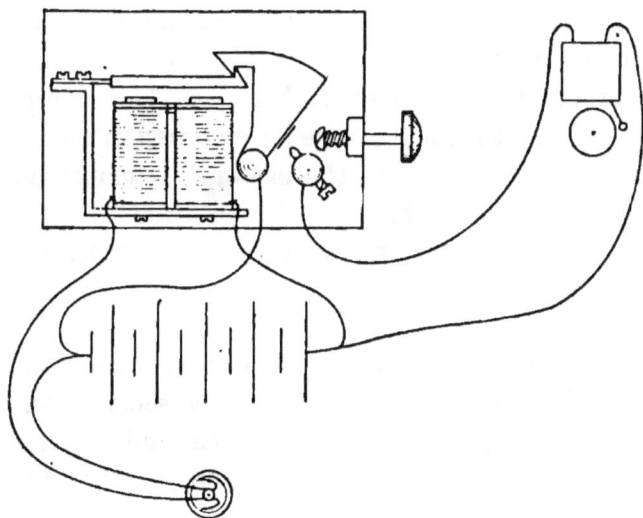

Fig. 50.—Continuous-action Bell, with Contact separate.

as already described; it is usual to take one main or battery wire directly from one pole of the battery, by the most convenient route, to the neighbourhood of each push, a short piece being jointed into it as near as possible, and carried direct to the push.

A second wire leaves each push, and is carried

L

directly to the indicator, so that there are as many wires leading to the indicator as there are numbers; and there are that number of wires and one more to carry down to the neighbourhood of the indicator, the extra one being the main battery wire. A short wire leads from a terminal on the indicator to the bell, which is usually fixed just above it, and in some cases forms part of it, and another wire from the other terminal of the bell to the battery, on the opposite side to the main wire.   .

In practice it is usual to bring a short length of double wire,—that is, two insulated wires in one covering,—from each push to the passage, or main channel where the wires are to run. There they are jointed, one to the main battery wire before mentioned, and the other to the wire that is to lead down to the indicator.

Or, if preferred, one short wire may lead from the push to the main battery wire, and a second wire be taken from the push right away to the indicator.

The double wire is more convenient for the branch circuits leading from push to push.

Where it is required to ring a bell, say, in the servants' bedroom, or in any other place away from the indicator, from a push in one of the rooms, a wire must be led to the bell it is required to operate, from the opposite side of the battery to the main wire; the same side as that leading to the indicator bell.

When there are many such connections, a second battery main is carried to the neighbourhood of all the rooms, with the first; and short wires are led from it to the bells or other apparatus.

For the continuous-action bell, three wires are required at the bell; viz., two from the battery direct, and one from the push that is to start the trip action. The two from the battery lead, one to the bell itself, and the other to the contact pillar which the trip lever falls on, the lever itself being connected to the other side of the bell, so that on falling it closes the circuit, as already described. The wire from the push is connected to the same side or terminal of the bell as the trip lever. In fact, the trip lever and its contact form simply an automatic push.

Gauges of Wires, Fixing, etc.—The size and insulation of wire to be used for domestic bells are very important matters, over which it is easy to make mistakes.

It should be remembered, in all work of this kind, that the effects of time and the surrounding conditions have to be taken into account, if permanent success is to be obtained.

A very small gauge of wire will give very good results, for instance—as long as it lasts; but a small wire is easily damaged in fixing; kinked, nicked with a knife in removing the covering, and parted; and then, being usually hidden, the cause is not easily discovered, and can sometimes only be repaired by fixing a new wire.

So, too, cotton and silk are very good insulators; but a wire covered only with silk or cotton, even

double or treble layers, would not last long except
under conditions where no covering at all would be
necessary. Cotton and silk both rot, especially where
moisture is present; the covering falls off; and you
may have two wires, supposed to be insulated, making
good electrical contact with each other and ringing
your bell.

Wet cotton and wet silk too, as already explained,
are not insulators, but bad conductors; and considering
the cross-section that will be exposed to the action of
the current in the case of two wires running together
through a large house, its resistance may be sufficiently
low to allow of a current passing that will be a serious
drain upon the battery, and that will gradually oxidize
one of the wires until it parts in some weak place.

The wires for domestic house bells, therefore; for
those leading from the pushes to the indicator, should
never be smaller than No. 22 gauge, and would be
very much better if No. 20. Even No. 18 would not be
too large; but No. 20 usually meets all requirements.

The main battery wires should never be less than
No. 20, and as much larger as you like. The covering
will depend upon the situation. If it is a new house,
and the wires are fixed before the plastering is done,
they should be run inside bell tube everywhere, and
the ends that are left out for connection to the pushes,
bell, etc., carefully wrapped, otherwise they will be
seriously damaged by the plasterer's tool.

The covering for wires inside bell tube may be two
layers of rubber laid on in opposite directions under

tension, with an outer covering consisting of two layers of cotton well varnished, or run through paraffin wax ; or with a close stout braid, well varnished or paraffined. This will stand well, provided the walls of the house are not wet. A new house is, of course, always wet when first built; but some houses unfortunately never dry, and for these rubber would hardly do, as rubber will not stand wet for long. Gutta percha should be used where it is anticipated that moisture will be generally present; but again gutta percha does not stand a dry atmosphere nor a warm atmosphere long, so that if there is reason to anticipate either of these, it is not wise to use gutta percha.

Callender's bituminous compound claims to withstand either heat or wet, and should therefore be suitable for either situation.

In fixing house bell-wires, great care must be taken not to damage the covering, especially in places where they are exposed to moisture. In drawing them into the bell tubes, the inside of the tube should be examined, to see that there is no burr in it; and the tube itself should not be too small.

At junctions of tubes with each other, and where they join larger tubes, care should be taken that the coverings of the wires are not cut in passing over the sharp edge of the tube, and that no portions of the wires are left exposed to the plasterer's tool.

Great care is also necessary in making joints. These should be as few as possible. Work from a whole coil rather than joint, if you can; as a joint is always a

possible source of failure. Where you are obliged,
remove the covering very carefully, so as not to
nick the wire in doing so; if possible, burn it off; in
either case, scrape the wire clean and make the joint
shown in Fig. 51, which is known as a bell-hanger's
twist. It must not be confounded with the bell-

Fig. 51.—Bell-hanger's Twist Joint.

hanger's *loop* joint, shown in Fig. 52. This is utterly
unsuitable for any kind of electrical work, as it may
or may not make firm contact, according to the conditions surrounding it. It is wiser also *not* to solder a
joint. For domestic bell work it is not necessary; the
close twist shown in Fig. 51, carefully made, does not
offer any resistance of moment to the small currents
used, and if it is carefully covered with a strip of

Fig. 52.—Bell-hanger's Loop Joint.

rubber laid on under tension, two or three layers deep,
it will not give any trouble. Sheet gutta percha, or
Chatterton's compound, are also used for covering
joints; but they are not so good as rubber carefully
laid on, if the latter be protected from moisture.
Stapling of wires should be avoided wherever it is

possible, as in driving the staple home, the covering of the wire is often damaged; and in the presence of moisture and the lime of the walls, you have a small galvanic battery, which in time will eat both your staple and your wire in two.

Especially should this precaution be observed with the main battery wires, as any action that can take place is always assisted by the presence of the current; though the resistance of the couple so formed may be so high that its measured strength is inappreciable.

Further, when it is necessary to staple house bell-wires, as in those cases where the wires are fixed outside the plaster; say in houses that are not new, or where the owner prefers having the wires in sight; care must be taken that no two wires are under the same staple if it is possible to avoid it, nor that two staples holding two separate wires are in contact. In either case, connection may be made between the wires, by the staple or staples, and a bell may ring or the battery be exhausted in consequence.

Fig. 53.—Staple for Domestic Bell Work.

Where staples are obliged to be used, the flat form shown in Fig. 53 is best, as they are less liable to damage the wire.

The author has been informed that some French houses use a small bone insulator, secured to the wall with a tack, the wire being strained around it.

Pushes.—The forms that a push may take are very

numerous. Any arrangement that can be made to
hold the two springs before referred to, in such a
manner that they are out of contact when the push
is not in use, and are easily brought into contact by
the pressure of the finger,—will answer for a push.
They are made in all kinds of wood, in ivory, in china,
in metal, in all sorts of patterns, carved wood, moulded

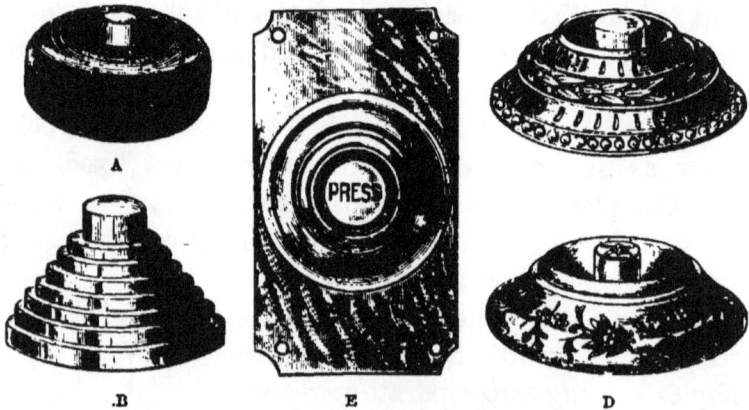

Fig. 54.—Forms of Pushes.

fancy material, ornamental castings, painted china,
glass, marble, serpentine, etc.

Where they are of metal, china, or glass, the back
is usually of wood or vulcanite, either screwing
directly into a thread cut in the inside of the top of
the push, or fitted to a metal flange that screws on
to the top. Figs. 54, A, B, C, D, E, show some forms of
pushes. It is far stronger, and more durable to have
a metal flange for the back of china, glass, and metal

pushes, than a plain wood or vulcanite back, as the thread on the latter is so very apt to strip, if the front has to be taken off two or three times ; and one always has to do this in fixing domestic electric bells, and during the time they are in use.

The pear push, one pattern of which is shown in Fig. 55, to which a silk rope is attached, is very useful for bedrooms. The rope is made long enough for the push to hang on the pillow, so that the bell it communicates with can be rung without getting out of bed. This is of very great service in case of sickness, or where servants have to be called in the morning. They have not been much adopted in hotels, from their being so frequently pulled down in the early days of electric bells, visitors mistaking them for the old form of bell-ropes, that a prejudice has naturally been excited against them. Probably they may be again introduced in those hotels where the comfort of visitors is studied.

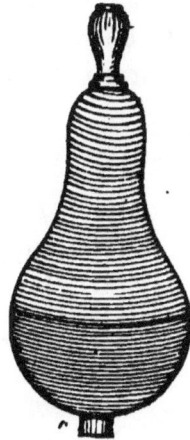

Fig. 55.—Outside of Pear Push.

The roof-box is a necessary adjunct to the pear push and rope, as otherwise, the weight of the rope and push would be on the joints to the leading wires, and the latter would be in danger of making contact with each other ; added to which, it is not easy to make a good joint between a strand of very small wires, such as are contained in the silk bell-rope and a single wire. The roof-box is merely a sort of push-box, without a push button or springs, but with two brass plates screwed firmly to its base, each plate having two flat-headed screws—one for the leading wire, and one for the strand of the rope. The front of the roof-box unscrews in order that the connections may be made. The rope is led through a hole either in the under side or in the front, the leading wires through the top, so that when fixed the apparatus presents a very neat appearance. One form is shown in detail in Fig. 56.

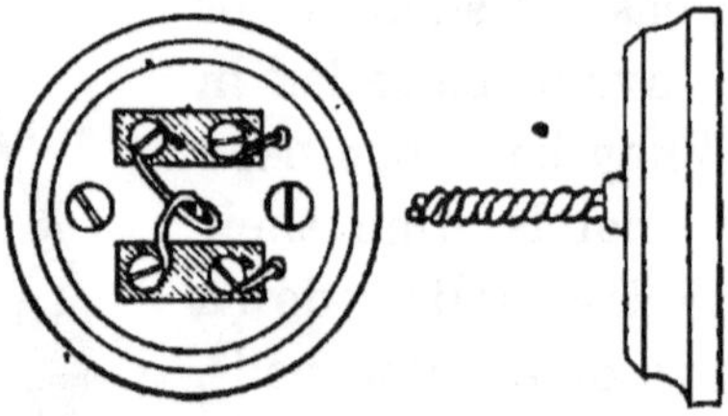

Fig. 56.—Roof-box.

It is necessary, of course, that the wood bases of pushes should be strong, of well-seasoned wood, and should not split when screwed up. The springs may be of the forms shown in Fig. 36, page 132; and they may either be screwed to the base with two small screws, and have a third flat-headed screw for the wire; or the screws that hold the spring may be made a little larger, and do duty as terminals for the wire as well. The former is by far the better plan, as it is easier to make a good connection, and there is less chance of the spring not being held securely in its place. The bases of pushes should all be drilled with separate holes for each connecting wire, large enough to allow the covered wire to pass through, so that the bare wire need not be exposed except where it makes connection with its own spring; two larger counter-sunk holes for the screws that secure it to the wall should also be provided.

Figs. 57, 58, 59, 60.—Pushes and Pulls for Front Doors.

With pear pushes, a single hole is usually drilled through the centre for the two strands, and some care is necessary to prevent the wires coming into contact in the hole.

For front or back doors, the contact may be made either by pushing in a centre piece, as with all the others, or by pulling out a knob as in the old pull bells,

In either case, some form of barrel is usually attached
to an ornamental bronze front, having the push or
knob in the centre. Inside the barrel works a spindle
or plunger against the tension of a strong spiral
spring. The office of the spiral spring is to keep the
two connections *out of contact,* until the pressure of
the finger, or the pull of the hand, brings them together. Figs. 57, 58, 59, 60, show forms of pushes
and pulls for front doors.
The best form of arrange-
ment for these out-door
contacts is a pair of stout
brass springs, held, in some
cases, inside the barrel
before mentioned, but in-
sulated from it, and from
the spindle or plunger. The
latter carries at its end an
insulated metal disc, which,
when the plunger is pressed
in or pulled out, as the case

Fig. 61.—Front-door Pull.

may be, makes contact with the brass springs on each
side of it, and so closes the circuit. On the hand
being removed, the spiral spring before referred to,
pushes back the insulated disc and breaks the circuit.
A pull on this plan is shown in Fig. 61.

Sometimes the plunger is made to form one contact
and the barrel the other, connection being made when
the plunger is pressed in.

Another form of front-door pull has been adopted

from the old pull bell. A brass plate, carrying a ring handle, is mounted on a marble, slate, or hard wood base, which is fixed in the usual position. To the ring handle is attached a stout bell crank, carrying a contact spring, and at its fulcrum is a clock spring, which winds up as the handle is pulled out. Attached to the plate, but insulated from it, is a contact point, so placed that the contact spring can rub over it, as the handle is pulled out; and, on the handle being released, it is pulled back by the spring, the contacts separating. This arrangement works remarkably well if properly fitted, as the rubbing contacts keep clean, and avoid the interposition of the resistance that sometimes arises from the dirt caused by the spark on breaking circuit.

One point that has to be carefully attended to in connection with front-door pushes and pulls is, the complete exclusion of moisture from the contacts. This is sometimes exceedingly difficult to accomplish, if the wires are exposed to the atmosphere on their way to the push, as in the case of an outer garden gate; or if any path is left for moisture to pass through at a front door.

The plan usually adopted for front or back doors, is to make a hole in the stone door-post considerably larger than the barrel before described, but not as large as the ornamental front of the push or pull. Into this hole a block of wood is driven, having a hole in its centre just large enough to admit the barrel. The latter is forced into the hole in the wood block, the wires led to their connecting screws, connections carefully made, the hole plastered over on each side, and the ornamental front fixed in its place.

All ,push springs, and bell contact springs, except those where a rubbing contact is formed, should be tipped with platinum. The object of this is, to prevent the contact surfaces being so readily dirtied by the small spark which passes whenever a circuit is broken. The spark so formed always carries over, from the positive spring to the negative, a minute quantity of the metal of which the contacts are composed; and it naturally carries over more of the less refractory metals, such as brass and iron, than of platinum. The minute quantity of metal carried over is deposited in the form of a fine powder on the negative contact, and gives the latter the appearance of being dirty. The fine powder, of course, introduces a resistance into the circuit, and so weakens the current; and it may even cause a bell to refuse to ring, if the layer of powder be very thick, or the battery or electro-magnet of the bell be weak. Platinum itself dirties, and even wears through in time, if it has much work, or is thinly put on.

Great care must be used in cleaning platinum contacts. It is usual to scrape them gently with the small blade of a pen-knife; but a better plan is to place a piece of clean glazed note-paper between the contacts, hold them together, and draw the paper gently through two or three times until the powder is all removed.

A form of contact that is very useful for connecting out-door garden or courtyard gates to the electric bell system of the house is shown at Fig. 62.

It consists of a pull contact, enclosed in a small box, and is worked usually by a crank bell-wire led from the gate, the contact box being in-side the house.

Inside the circular box are usually two springs, which represent the two sides of the circuit, just as in the ordinary push or pull. A brass lever, which is worked by the crank bell-wire, makes connection between these two springs, and so closes the circuit.

A spiral spring, whose tension has to be overcome by the crank bell-wire, pulls the lever back in its place after the pull is released.

These pulls may also be used for bedrooms, as shown in the Figure 62, the old-fashioned silk ropes being used to actuate them. It must always be remembered, however, in using them, whether at an outer door or in the house, that the old plan of pulling and releasing with a jerk will not answer with electric bells. The pull or rope should be pulled down or out, and held there for a few seconds, and then

Fig. 62.—Pull Contact for Outside Gates or Bedrooms.

gently released. Pulling it down or out violently, and letting it go with a jerk, may damage the spiral spring in course of time. Any way, it is putting the apparatus to an unnecessary strain.

Fire and Thief Alarms.—A very useful extension of the Electric Bell system is its adaptation for giving warning of fire, or of any one entering a house at night.

For either purpose, all that is required is a contact similar to a push in principle, that shall be actuated by heat or by some action of the burglar, such as opening a door or window.

It is obvious from what has been said on the matter of branch circuits, that the simplest arrangement, as well as the cheapest, would be to carry a branch circuit from the wires leading to the pushes in each room to the fire and burglar alarms respectively; and to complete the arrangement, a large bell could be placed in a convenient position, where it can be heard by every one, the burglar included, and switched on at night in place of the indicator and bell used for the regular service. The alarm bell could be placed in the gardener's lodge, or even the police station.

In some cases, however, it would be necessary that the room where the fire arose, or where the burglar was, should be indicated to the master of the house. When that is so, a separate set of wires are used, a separate indicator being fixed where desired, and a separate battery; the automatic circuit closers taking the place of the pushes as before.

In the writer's opinion, fire alarms might be used much more than they are. It is now pretty well established that the first quarter of an hour is the all-important time in cases of fire. When a fire is detected early, it may generally be extinguished without serious damage having taken place; whereas, if it be allowed to gather strength, the damage it may do is only limited by the state of the atmosphere and the materials that lie in its path.

In America, where many of the buildings of the younger towns are of wood, the automatic fire-alarm system has been brought to a great state of perfection; engines being out and on their way to the scene of the fire within a few minutes of the temperature reaching a certain degree.

Neither fire nor burglar alarms have been used much in this country; and the Author believes that the reasons are mainly twofold.

In the first place, neither fires nor burglars are of every-day occurrence; and the British householder naturally grudges the expense attached to any apparatus that he does not consider necessary, and that does not add to his convenience, no matter how beautiful the apparatus may be in the abstract.

And secondly, neither fire nor burglar alarms have always been successful. In any case, it is necessary that they should be tested periodically, and that the average British householder would have a decided objection to; though for large public buildings, for banks, theatres, etc., the trouble should be well repaid.

M

A description of the various forms of alarms in use will reveal the causes of their failure.

One favourite plan is, to have two pieces of metal, insulated from each other, connected to the two wires representing the opposite ends of the circuit; and held apart by the pressure of the door or window when shut, or by some substance that melts or elongates under the influence of heat. In each case, some form of spring is held in tension the whole time that the alarm is *not* acting, and only comes into play when it does act; the consequence being, that the spring is very liable to be strained, and to refuse to act when required to do so. Straight springs in particular, when held back for long, are liable to remain so. Also where moisture can enter, steel springs and screws are liable to rust, and even brass springs are liable to become corroded, especially if there be any leakage on the circuit.

Fig. 63.—Plunger Door Contact.

The best form of contact for doors is the one shown in Fig. 63, where the closing of the door forces in a small plunger, in opposition to the tension of a spiral spring, thereby breaking the electric alarm circuit; on the door being opened, the spiral spring forces the plunger back into its place, remakes contact, closes the circuit, and rings the bell. A small brass plate should be fixed to the door opposite the plunger, otherwise the latter will wear a hole in the wood, and will then not be forced out of contact when the door closes.

For windows, the form shown in Fig. 64 may be used; the frame of the window sliding over the bevelled button of the contact, and forcing it out, as it comes down; the reverse taking place as the window is pushed up.

These alarms are very useful as calls for shop doors. In some cases the door is made to work a contact during a portion of its sweep in opening, so that the bell rings for a few seconds, and then stops; ringing again in the same way as the door is closed.

In other cases, the bell is made to ring as long as the door is open, unless when switched off by the attendant, with a switch provided for the purpose behind the counter.

Another form of contact that is sometimes used in shops, and that might be used anywhere, is the floor

Fig. 64.—Window Contact.

contact. The floor, under the door-mat, is cut away, and suspended on springs. One or more contacts of the form shown in Fig. 65, consisting of a plunger with a flat top, are placed in a convenient position under the board. Any one walking over the mat, depresses the board, and with it the plunger of the floor contact, closing the circuit and ringing the bell.

Great care is necessary in fitting door and window

contacts, or any other form of automatic arrangement. The apparatus must not come into operation when it is not wanted. The householder will soon vote the whole thing a nuisance, if his family are aroused in the middle of the night for nothing. The changes of temperature, and the varying fitting of doors and windows consequent thereon, must neither cause the alarms to refuse to work when required, nor to work when they are not wanted to. It must not be possible,

for instance, for the shrinking of a window or door frame, in cold weather, to give sufficient clearance for the spring to make contact; nor for the warping of the outer frame of door or window to cant the moving portion of the apparatus, so that it cannot find its mate.

Further, all the wires should be out of sight, and out of reach of scientific burglars. It is stated that one of the strong rooms belonging to a safe company in New York, is fitted with two metallic linings, insulated from each other by a thin layer of paraffin paper, or other material; so that, should the scientific burglar put his chisel through, with a view of cutting the wires, he will make connection between them and ring the alarm.

Fig. 65.—Plunger Floor Contact.

The best form of electric fire alarm consists of two springs or plates of different metals, such as brass or steel, insulated from each other. When at normal

temperature they are not in contact. On a rise of
temperature occurring in their neighbourhood, one
spring expands more than the other; and so, one end
of each being fixed, the other ends come into contact
through the greater curvature of one of them, owing
to their different rate of expansion. In some, the
curvature given to a composite metal bar when heated,

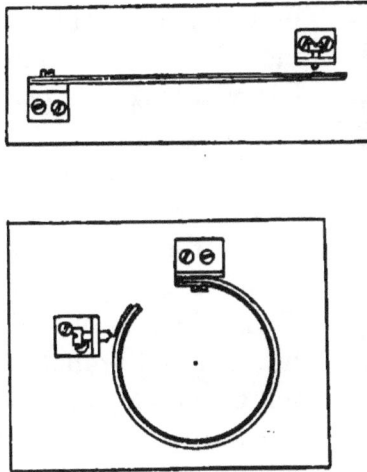

Fig. 66.

brings its free end against a contact piece, as shown
in Fig. 66. In other forms the contact pieces are kept
clear of each other by a substance which melts at a
certain heat, and allows a spring to pull them together.

In Fig. 67, a thermometer has been adapted to the
work, platinum wires being sealed into the bulb, and
into the stem, at a point opposite the degree of heat

at which the alarm is to come into action; or into the top, the wire being allowed to project down the tube to the given point. This would be the simplest form, if it were reliable; but the writer understands that there is a difficulty in getting the mercury to leave the second wire, after it has once been forced up to it. The contact is made, of course, by the column of mercury connecting the two platinum wires, the latter taking the place of the push springs.

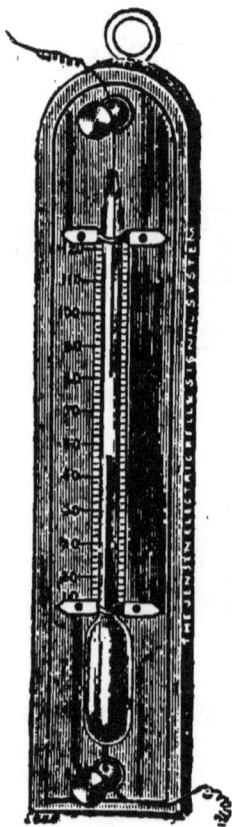

Fig. 67.

The proper designing of fire-alarms suitable for each case, is a matter requiring considerable thought and practical experiment. That which is suitable for a private house would probably be unsuitable for a theatre, and both would be unsuitable for the hold of a ship. Again, the apparatus that would answer for a cargo of cotton might not for a cargo of coal.

The Author believes that if this branch of the subject be properly attacked, many lives and much property will be saved.

Faults, or Causes of Failure.—The causes of failure of Domestic Bells are usually very simple, very easily found, and very easily repaired. First, suppose that none of the bells in a house will ring. Clearly the cause must be in something that is common to all the bells. Either the battery, the bell itself, the indicator, or a wire leading to all the pushes, must be at fault.

In such a case, always begin with the battery ; or in fact, in any case, when proceeding to look for faults, begin with the generator, whatever it may be.

Electrical Engineers, when called in to put some apparatus right that is not working, would always do well to spend a few minutes in finding out exactly how it has failed—what it will not do. The user generally will only say at first, that the thing will not work, and that it is a great nuisance ; but a few judicious questions will extract all the information required, and save a great deal of trouble.

First then, test the battery, as already described, page 117, Fig. 30. If that is all right, carefully examine the indicator and bell, and look for broken wires. The ends of wires that are scraped bright, and clamped under terminal screws, become brittle in time, and may be broken off by the flick of the housemaid's duster. Occasionally, too, the wire itself will be broken, and yet remain in position, held by a portion of its covering, or by the coil that is usually made at the end of a wire, to allow for these breakages; so that a close scrutiny and a somewhat practised eye is necessary to detect the breakage. Also, if there are indications of damp anywhere on the walls, carefully examine the wires in its vicinity. If the covering of the wire is thin, moisture may have worked through, attacked the conductor, and with the assistance of the current, or some flaw in the wire, parted it inside the covering. The surest sign in these cases is the green oxide of copper,—verdigris,—which always forms. Wherever this is visible, trouble will follow.

If nothing can be discovered, take a piece of covered wire, connect it to the same end of the battery to which the main wire leading to all the pushes is attached, and connect it to the following in succession :—

Each terminal of the indicator.

The terminal leading from indicator to bell.

The bell terminal on the indicator side.

A rule that may as well be stated here for finding faults, such as broken wires, etc., is: Connect the faulty wire or apparatus with the generator, in this case the battery, so as to have a difference of electrical potential between two parts of the circuit, that can easily be got at. Connect these, at successive points, by a wire, in which is inserted some apparatus, such as a galvanometer or bell, that will show the presence of a current; working outwards from the generator. When you find two points, or pairs of points, between which a sudden change occurs; such as, at one point you have evidence of a strong current, and at the other either no current at all or a very much weaker one, the fault lies between those two points. In the present case, your battery is connected to the two wires leading to the pushes; on one side being the main battery wire, and on the other the indicator and line wires; so that the required difference of potential is present. Connecting one end of a wire that you know is sound,—and in which you may insert, if you like, a detector galvanometer; though this is not usually necessary, the indicator and bell answering the same purpose,—to the battery terminal from which the main wire issues, and touching its other end successively on the points named; if the fault be in either the indicator or the bell, the failure of these to work, and there being no deflection of the detector galvanometer, if one is used, will denote the fact.

It will easily be understood that when you connect your testing-wire to the various points named, you close the circuit through different parts of the apparatus. Thus, when you connect to the indicator, you have a circuit comprising battery, bell, and indicator, and the wires from the battery between bell and indicator only. When you connect to the bell terminal of the indicator, you have the bell, and the wires from battery to bell, and from bell to indicator only in. As you pass on, you cut out, first the wire connecting indicator to bell, then the bell itself.

It will be obvious also, that if the bell refuses to ring when you connect to the line terminals of the indicator, —those to which the wires leading from the pushes are connected,—and it rings when you connect to the bell terminal of the indicator, the fault must be in the indicator itself.

A little thought is sometimes necessary here, to avoid being misled, as we may go hunting for a broken wire inside the indicator, pulling the more or less delicate mechanism about, when it may be that the battery is weak, though it has not actually ceased to furnish a current. The bell may ring by itself, and refuse to ring with the resistance of the indicator coils added to the circuit, especially if either bell or battery be weakly constructed, without any reserve of power, and the indicator coils be of high resistance.

A galvanometer comes in here. If you have a galvanometer in the circuit, you can tell at once if you have a total break or not.

If you find a total break, or a great diminution of current strength with either the bell or the indicator bell only in circuit, proceed to test through the apparatus at each point where a connection is made till you find the two points as before. The cause in either apparatus may be a wire eaten in two at the back of the bell or indicator, owing to its being placed on a damp wall and not properly protected. The ends of coil wires are often taken through the backboard of an apparatus, sunk in a groove there, and brought through again at the terminals, or soldered to the latter underneath. This arrangement gives a neat appearance, but may prove troublesome, unless the wires are well protected with either paraffin wax or some other suitable substance. Where wires are soldered to terminals, if the wires are small, there is danger of a fault later on, unless the soldering is very carefully done.

Another rather common fault is, the inside ends of

coils of wire break off short just where they are coiled up to connect. The brittleness before referred to will account for this, and it may be also caused by vibration, or by the wires having been sharply bent over the end of the bobbin on which it is wound. In any case, the coil must be stripped and re-wound.

Other sources of trouble, which have already been mentioned, are faulty contacts in the bells, springs or screws working back, bell-domes working loose and away from their hammers, and dirt between contact points.

If neither the battery, the bell, nor the indicator are at fault, it must either be the main wire leading to all the pushes, or the pushes or their line wires themselves. It is more probable that the former is the case ; but it can be ascertained for certain by making the test-wire long enough to reach to one or more of the pushes, and testing by connecting to each in succession. If the bell now rings and the indicator works, it will be clear proof that the main battery wire is at fault. It is probably parted. If the wires are in sight, the testing may be carried further by bridging over each portion of the main wire in succession until two points are found, at one of which, when the test-wire is connected to it, the bells ring as they ought, and at the other they do not. The piece of wire between these will require to be renewed, unless a more searching examination reveals the actual fault.

Where the wires are concealed inside the plaster, a new wire must be fixed. It can be carried outside and round the skirting boards, etc., to the nearest push, when probably the connection between all the pushes will be sufficient for the rest. There is no difficulty in doing this, and a careful workman will so arrange matters that the new wire is not unsightly, even if it is visible.

If the complaint be, that one or more of the rooms cannot communicate with the kitchen, or wherever the indicator may be placed, the fault, if not in the indicator, may be either in the pushes themselves or in one of the wires leading from them. If the rooms which are " off " are supplied by one branch of the main battery wire, most probably that is at fault.

To make sure, however, carefully examine all the pushes, removing the fronts; and if a galvanometer be available, test for the presence of a current by connecting the galvanometer wires to the wires connected to the push springs, and compare the deflection, if any, with that given at a push which is not " off."

The long and short circuits of the detector galvanometer are very useful for this work. Usually, when everything is in order, a moderate deflection is obtained on the short or thick wire circuit. When the battery is weak, and things not quite in order, no deflection may often be obtained on this circuit, leading to the erroneous conclusion that there is an actual break. The long circuit, however, will show a deflection as long as there is no actual break; and it is therefore possible, by testing with both circuits, and comparing the deflections obtained with former tests, to form a fairly accurate judgment as to the cause of failure.

It will sometimes happen in the case of pushes, as of wires, where they have been fixed upon damp walls, that the damp has found its way into the push, destroying the connection by the layer of oxide or chloride of copper formed. Sometimes the effect will be only to weaken the current by the extra resistance offered by the film of oxide; in other cases the wire may be eaten quite in two. More frequently, where a fault occurs at a push, it will be due to the wire having broken off short where it is clamped under its screw; and it will often cause considerable annoyance by sometimes touching its spring, allowing the bell to ring, and then a little later, perhaps owing to some vibration in the house, say, due to a passing heavy vehicle, the connection will be broken, and it will fail.

Either of these faults, breakage or oxidation, are visible immediately the push is examined; and the latter generally gives warning, by its bell ringing weaker and weaker as the oxidation progresses. Occasionally, where push springs are not tipped with platinum, the dirt formed by the spark will cause trouble, as before explained; but this is usually where other parts of the system are weak. If the battery and bell are as powerful as they ought to be, the ringing rarely stops from that cause.

A not infrequent fault with the pear pushes and silk ropes that have been described, and a somewhat troublesome one, is a break in the flexible wire inside the cord. In order that it may be very flexible, these are made of a number of very fine wires stranded together and then covered with silk; the completed silk strands being laid up together into an elegant rope.

It may happen that the rope is a trifle too long, and is shortened by a knot being tied in it. After a time, the sharp kink formed by the knot parts one after another of the small wires in the strand, till all are broken, and the bell fails, but in a most exasperating manner. Sometimes it will ring, and the next minute fail to do so. The reason is, the broken ends of the wires, lying very close together, come into contact when the rope is in one position, say curled up, and do not make contact when the position is altered by some one having used the push to ring and let it fall.

To find out if the fault is in the rope, in these cases; test at the push springs, and again at the roof box, where occasionally similar faults may occur to those described in pushes. If there is much difference in the deflection of the galvanometer needle at these two places; or if the bell rings when a short piece of wire connects the two terminal plates of the roof box, and does not when the ends of the flexible wires are connected, remove the push and carefully examine the rope, looking for a place where one of the strands is not so solid as the others. The break can often be detected in this way by an experienced hand. In any case, the rope must be cut off at the break, as it cannot be repaired. If, when the damaged piece is removed, the rope is too short for its purpose, a new one must be fixed. If the break cannot be found by a careful inspection of the rope, it can with certainty by cutting off successive portions, connecting the ends of the wires each time, until the bell rings again, when the fault will have been cut off.

If the fault is not in the pushes, bridge over successive wires. First, if several pushes are off, bridge over the main wire, by leading a sound piece of wire from a push that does not ring to one that does. If the bell now rings, it is the main wire which is at fault, and this may be replaced by leading a new wire from the pushes that have failed to any convenient one where the main is still good, fixing it as described above.

If it proves not to be the main, the fault can only be in the line wire, leading from the push to the indicator. Bridge this over temporarily to test, as already described, and then permanently, as with the main wire.

Faults in special bells, such as from servants' room to master's bedroom, etc., are dealt with in exactly the same manner, remembering that the break, if it be in a wire, may be in the main battery wire to the push, the line wire from the push to the bell, or the return battery wire from the bell to the other side of the battery.

CHAPTER VI.

SIGNALLING by electricity in mines was first introduced in Yorkshire, and in one colliery in South Wales, Cwmaman, not long after the introduction of electric bells into domestic life. The early signals were at great disadvantage owing to the want of a simple battery that would stand the hard work that was required of it ; and even when the Le Clanché battery was introduced, it was so little understood that it was speedily condemned for mining work. The small Le Clanché cells that were used would not stand the frequent short circuits, and other drains upon it that were common in mining work, so recourse was had to the old Daniell, the only battery then known that would keep up to its work. When the Author first took up mining work, in 1876, he was handed over a battery, the invention of his predecessor, that apparently stood short circuiting better than the Le Clanché, but was gradually displaced by the Le Clanché, as the valuable qualities of the latter became better understood.

Another grave mistake that was made in the earlier signals was : everything was too weak. In some cases the bells that were not doing very well in private houses, were thought good enough for collieries. Small wires, badly insulated; small battery cells; weakly constructed ringing keys, with no provision against damp or rough usage, were generally used; and neither the contractors who fixed the signals nor the officials of the mine understood the work as they ought. Moreover, the former frequently endeavoured to keep what little knowledge they did possess entirely to themselves, even going so far as to lock up the battery boxes, and refusing to assist the officials in any way to become masters of the apparatus, for fear it should lead to their losing business. They also, in many instances, charged prohibitive prices.

In all mining work, two things are necessary before all others; viz., strength and simplicity; and no matter how beautiful an apparatus may be, it should stand condemned unless it fulfils these conditions.

The bells should be made large, all their parts strong, with proper provision against changes of temperature, vibration, etc., and the electro-magnets should be very large, so that there may be a large reserve of power, as already explained in the chapter on Electric Bells.

Bells and Keys.—Six inch, eight inch, and eleven inch bells are generally used for mining signals, and more frequently single-stroke bells than tremblers, as they give a clearer signal, and do not get out of order so easily. The electro-magnets of the 6 in. single-stroke mining bells made by the Author's firm are turned out of 2 in. round iron, and the 8 in. out of $2\frac{1}{2}$ in., everything else being in proportion. The earlier bells were made with ¾ in. iron cores for 6 in., and 1 in. for 8 in. bells, stout brass collars being used for coil ends. Though some of these bells have been in use twelve and fourteen years, there has not been a single instance of one coming in for repair.

It is also not wise to make the resistance of mining bells too low; as, being subject to such frequent changes and short circuits, the battery would be very much more severely strained than where the resistance is moderately high. As, too, a higher resistance means finer wire and more turns, the proper working current, being smaller, is more easily maintained than where the turns are fewer and the resistance lower.

Ringing keys or contact makers form a very important feature in mining signals. They also require to be made very strong. Except for dry roads, cabins, or engine houses, wood mountings will not answer at all; and it is questionable if iron would not be better everywhere. All connections between the wires leading from bells or batteries to the keys should be in sight, and easily to be got at; the reason being, that where damp is present, it always forms a film on the outside of the wire or its covering, and follows the latter to its destination, the terminal screw. As it cannot be covered there, it always proceeds to attack both wire and terminal, forming the green oxide of copper already referred to, and in time destroying the connection. If the connection to the terminal screw is hidden, as in some forms of keys, the action goes on all the same, only that no warning is given, and the signal suddenly fails with nothing to show the reason; whereas if the terminals are in sight, the green oxide is visible and can easily be cleaned off from time to time before it can do much real harm.

Figs. 68 and 69 show the Mining Key made by the Author's firm for shaft signals.

It consists of a metal box made in three parts, the joints being dovetailed together. One part forms

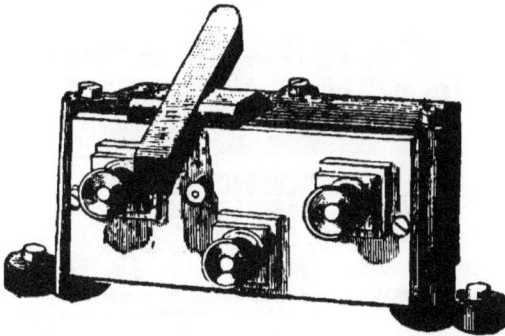

Fig. 68.—Outside View of Shaft Key for Mining Signals.

the body of the box, and is of iron, having three lugs for the screws which hold it in its place. The lugs are arranged so that the key stands away from its support, allowing any water that may be present to run down behind.

The front of the key is also of iron, and carries a pair of lugs which form a hinge for the iron lever which is shown. The front is secured to the body of the case by three screws, so that it can be removed,

and the working parts examined without stopping the
signal.

The bottom is of brass, and carries the working
parts of the key and the terminals. The working
parts of a bent brass lever, pivoted as shown in Fig.
69, held by a strong spiral spring at one end, and
working between two stops at the other. The stops
and the fulcrum of the lever are mounted on a vulcanite standard secured to the base.

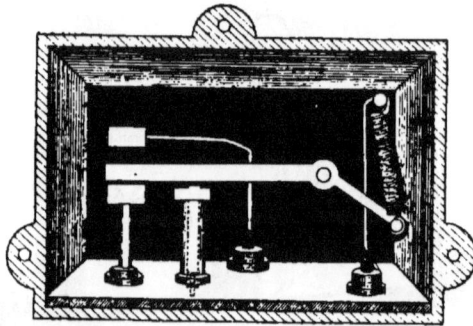

Fig. 69.—Inside View of Walker and Olliver's Shaft Mining Key.

A brass pin passes through a hole in the base of
the key and through a brass guide attached to the
vulcanite standard, and it is this which works the
internal brass lever, being itself actuated by the iron
lever pivoted on the front of the box. The terminals
are all attached to the base of the key, their connecting screws projecting below it and their connections
to the lever being inside the box. The terminals are,
of course, insulated from the brass base piece.

It will be seen that the above arrangement is very strong, and that damp cannot easily penetrate to the inside of the key.

Where the key performs the double office of transmitting signals and receiving them, the brass lever rests upon a stop which is connected to a terminal leading to the bell. The lever itself is connected to the terminal to which the line wire is attached, and the upper stop with which the lever makes contact when forced up is connected with a third terminal leading to the battery.

Where the key is only required to close the circuit, and so transmit a signal, and is not required to receive one, the lower stop and its terminal are omitted.

Fig. 70 shows one form of plunger key, that is used some-

Fig. 70.—Plunger Key.

times on colliery banks where it is not much exposed to the wet.

It consists of a dome or cone, mounted on a very stout board. The dome carries a plunger in the centre, something similar to that used for front-door bells, but stronger. It is kept back by a strong spiral spring when not in use. The board carries a spring working either between two stops, or simply facing another stop, according as it is required for single or double work; for transmitting and receiving, or only for transmitting.

The connections to the terminals are sunk in the face of the board, and the terminals themselves are placed on the face; the object being, as before, to keep them and their connecting wires in sight, so that any of the green verdigris that forms can be removed.

This key more nearly fits the collier's ideas when the knob of the plunger is made large, as he can hit it a blow with his fist, or bring all his weight to bear on it behind the palm of his hand; but it is unsuitable for pit bottoms, as the board itself would get saturated with wet, and then form a leakage path for the current. The dome of this key is sometimes made in brass, but more frequently in iron; when made in brass, it is usually mounted on a polished mahogany board, the connections being led through the base by pieces of insulated wire, jointed to the leading wires outside.

Fig. 71.—Cornish Key.

This form has been used for shaft signals, and on roads; but the Author understands that it has always given trouble where damp is present.

The joints outside are also objectionable, as they are possible sources of failure.

Fig. 71 shows a form of key, intended to close the circuit and transmit signals only, that has been designed by the Author for Cornish and lead mines, where a number of keys are required, and where it is very wet.

Its form is similar to that last described, but the dome and the base are in one casting; the object being to prevent water getting to the contacts, which, as will easily be understood, are the vital points of all circuit closers. In this key the plunger forms one side

Fig. 72.—Wood Base Key, for Dry Roads, or Surface Work.

of the circuit, being insulated from the dome by a vulcanite collar; and it makes contact with a stop at the bottom of the base when forced in. One terminal is on the outside of the casting, and the other is attached to the plunger collar, both being in sight as usual.

Fig. 72 shows a form of ringing key that is used for dry roads, cabins underground, etc., where only a circuit closer is required. An oak board has a stout straight brass spring secured to it by one end, the other end being fitted with a thumb-piece above and

a contact-piece below.   Fixed to the board opposite
the contact-piece on the spring, is a second contact-
piece, as shown.   When not in use, the spring is held
down by a bridge at a convenient distance from the
lower contact.   A wood cover and two terminals for
connection, complete the arrangement.   One terminal
is connected to the spring, and the other to the lower
contact, and the circuit is closed by pressing the
spring down, till the upper contact touches the lower
one.

Fig. 73.—Inside of Tapper Bell; *i.e.* Bell with Key attached.

Recently the Author has modified this key by using
an iron case, in place of a wooden one.

The case is made in two parts as before, the spring
being secured to the iron base, and the case itself
forming part of the circuit.   Where this is an inconvenience, the spring is insulated from the case by
means of vulcanite.   The object in this design has
been, to obtain greater strength.

Fig. 73, shows what is known as a tapper bell; that
is to say, a bell with a ringing key attached.   It is
used for surface signals generally, and sometimes,

especially in the North of England, for underground, where it is required to ring from station to station.

The bell is constructed as before, except that the electro-magnet coils are usually smaller, since they are not subject to such hard work as the others.

The key is adapted for sending a signal and receiving one. It may be either a lever pivoted near the centre, having a contact at each end, or it may be constructed similarly to the key shown in Fig. 72, page 183; in this form, the bridge against which the spring rests when not in use, makes the back contact. The back contact is connected to the bell coils, the front one to the battery terminal.

The connecting wires are usually under the wood base of the bell case, which is made hollow for that purpose, and three terminals are attached to one end of the base—for line, battery, and earth.

In mine signals the line wire is the one that connects the key at one station to the key at another, or to the bell, if there is no key. It is always carefully insulated.

The battery wire is the wire leading from one pole of the battery to the ringing key, or to the bell, where there is no key. This is always *very carefully* insulated.

The earth, or return wire, is the wire leading to earth or ground, where that is used, or that connects together the ends of the battery and of the bell coils which are not connected to the key. It forms a return path for all circuits. It need not be insulated, except where there is a difficulty in obtaining good insulation on the line wire.

The signals at present in use in and about mines are :—

The Engine-plane Signal.

The Shaft Signal, and

The Surface Tapper Signal.

With the Engine-plane Signal, of which many more have been fixed than of any other form, a bell and battery are placed in the engine-house, the bell as near as convenient to the engine-man, and the battery where

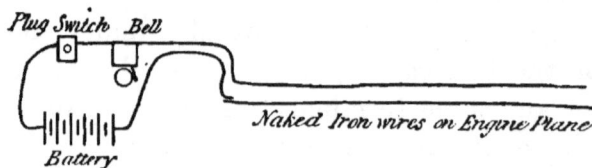

Fig. 74.—Diagram of Connections of Two-wire Engine-plane Signal, with one Bell.

it may have the advantage of a medium temperature at all times ; or, if this is not obtainable, in the situation which most nearly approaches to this condition, and which is also dry. Connected to these by means of insulated copper wires are two naked iron wires, which run side by side for the full length of the engine plane. The connections are arranged as shown in Fig. 74, so that by forming clean metallic contact between the iron wires, the circuit is closed, and the bell rings.

It is manifest that such a signal, if reliable, must be

of immense value in mining work. In many cases, it is perhaps not too much to say that it has secured the success of a given system of mechanical haulage.

It usually happens, in coal mines particularly, that haulage roads, when first laid out, wind a good deal, the hewers having followed the path of least resistance. Often very sharp curves are encountered; and in the long wall system of coal getting, as worked in some collieries in Derbyshire the junctions are so many that it has been found absolutely impossible to successfully operate a mechanical signal. Again, it frequently happens that the hauling engine is on the surface, the rope being led down the shaft, and great difficulty is experienced in communicating quickly with the engine-man.

To the electric signal these cases present no difficulty whatever, provided that it be properly constructed. No multiplicity of junctions or of curves make any difference in its efficiency; and it will ring its bell as easily on top of the deepest shaft, from the extremity of the longest road, as from 100 yards distance.

Where the engine is on the surface, a second bell may be inserted in the circuit at the pit bottom, to ring with the engine-house bell, so that the man at the pit bottom will be informed of all that is going on. Fig. 75 shows this. It will be wise, where this arrangement is used, to add a few cells to the battery, to overcome the additional resistance of the bell, and to provide for the additional work the battery has to do,

This is not absolutely necessary; as, if the bell and battery have been properly arranged in the first instance, the insertion of the second bell will make no apparent difference; but it will make a difference in the cost of maintenance, and in the attendance required to the battery, as the work done upon the second bell is taken from the reserve of power in the battery.

Another modification that is often arranged, is the addition of a bell at the far end; or in-bye end of the

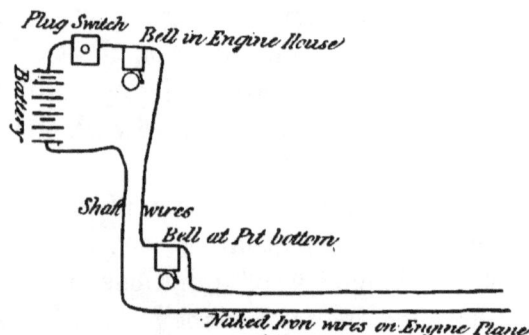

Fig. 75.—Diagram of Two-wire Engine-plane Signal, with the Engine at Bank, and two Bells.

road, as it is called in the North of England; to ring simultaneously with the engine-house and pit-bottom bells, where there is one of the latter.

To accomplish this, the same two iron wires are used, and earth or ground, at the battery and in-bye. Fig. 76 shows the connections for this arrangement.

As it is sometimes difficult to obtain good earth at the end of an engine plane,—that is to say, as it is often difficult to obtain a good conducting path for the current, through the crust of the earth in which the mine is formed,—a third wire is sometimes used in its place. This wire need not be insulated, but should be stapled well back out of the way. It is always better to use earth if possible, as the addition of the third wire destroys the simplicity which is the great desideratum in mining work. If the mine is at all damp, there is usually not much difficulty about the matter; and even in very dry mines, if proper arrangements be made to secure a large surface of contact

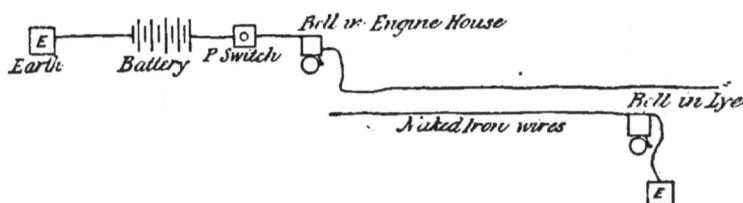

Fig. 76.—Diagram of Two-wire Engine-plane Signal, with a
second Bell In-bye, using Earth for Return.

with the body of the coal; almost the only exception would be, where the coal seam is very thin and its surroundings hard, dry rock.

On the surface, or near the engine-house, there is rarely any difficulty in finding earth; but it must once more be pointed out, that here too, a compliance with the conditions imposed by the laws of resistance is necessary. The end of a wire touching the ground will not make good earth, because the contact resistance between the wire and ground will be too high.

Another modification of the engine-plane signal is shown in Figs. 77 and 78. It is the one generally used with endless ropes and it consists of a number of bells at different parts of the engine-plane, usually at junctions, in addition to the regular ones at the engine-house and in-bye.

This signal may be worked either by " earth " at each station, as shown, or by a return wire, the former being preferable for convenience, the latter for safety, as the multiplication of earth connections is rather apt to lead to trouble.

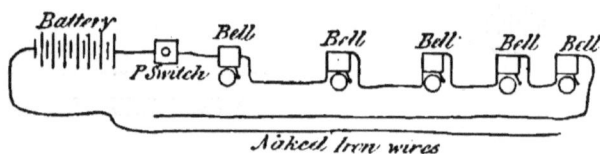

Fig. 77.—Diagram showing an Engine-plane Signal, with four Bells ringing together, the Bells connected in Series.

The bells may also be arranged either in series, as Fig. 77, or in parallel, as Fig. 78. The former makes the least drain on the battery, but demands the largest number of cells; the latter is more convenient and less liable to get out of order, if the battery and bells be properly constructed.

The great danger in using " earth " always is, the greater possibility of leakage; and for that reason, if the cells are too small, the parallel-earth system very quickly breaks down, in a damp mine.

The bells used for engine-plane signals are, 6 in. or

8 in.; in a few cases, where engines are very noisy, 11 in.; sometimes single-stroke and sometimes trembler, more frequently the former, though the latter has many favourites.

Whether 6 in. or 8 in. is a matter depending upon the money to be spent and the situation; but the 8 in. if properly made, is usually worth more than the difference in cost between it and the 6 in.; and, on the principles already so frequently detailed, will save the increased cost in decreased maintenance.

The iron wire used is ordinary telegraph wire, well

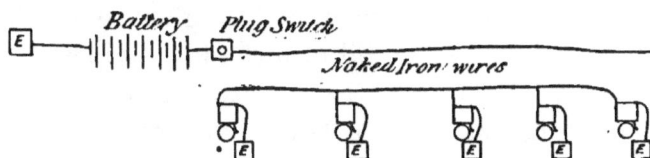

Fig. 78.—Diagram showing Engine-plane Signal, with four Bells ringing together, the Bells connected in Parallel.

galvanized, either No. 11 or No. 8. The latter makes the neatest job, stands best, and gives an increased margin of reserve, by reason of its lower resistance. One of the greatest difficulties that engine-plane signals have to contend with, is breakage from falls of roof, leading to numerous joints, usually not too well made.

These joints offer considerable resistance, owing partly to the dirt, oxide, etc., that is present between the opposing surfaces of the wires, but more to the bad contact that is made between them. Joints in

naked iron wires should be made either as Fig. 79, the
lapped joint, or the twist joint already shown for small
wire in Fig. 51, page 150.   The ends should be scraped
clean, in every case, before making the joint; and,
whether a twist or lapped joint is made, the two surfaces should be so firmly in contact as not to offer

Fig. 79.—Lapped Joint.

much resistance.   Under the very best conditions it
will offer a resistance considerably higher than that of
a single wire, and every bit of dirt adds.   Miners' joints
are usually made with two loops hooked into each
other, as Fig. 80.   When the wires are pulled very
tight, these joints offer a moderately high resistance ;
but if from any cause they become slack, the resistance
may be very great indeed.

Fig. 80.—Looped Joint, as usually made in Mines.

It is for this reason that the Author has so strongly
recommended a margin of power everywhere.   Where
either battery or bell are made weak, or even with
a moderate reserve, such as would do for most work,
the first miner's joint stops the working of the signal,
owing to the large resistance offered having reduced

the working current below that at which the bell will ring.

In one case, at Linby colliery in Nottinghamshire, all the signals underground were constructed with galvanized strand wire, 7 No. 16's. This, of course, added considerably to the cost, but it very materially increased the efficiency of the signals. Breakages of wires were very rare, and the reserve power was high. It is right to say, however, that falls of roof are not common at that colliery.

Another modification, carried out at the request of the manager of Linby Colliery, was an arrangement to work all the signals from one battery, which was made very large, and placed on shelves in a cabin near the pit bottom. Each signal was fitted with the plug switch shown in Fig. 81 ; and in case of anything happening on one road,

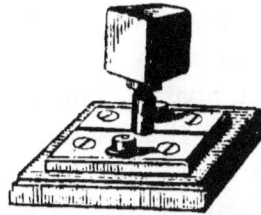

Fig. 81.—Plug Switch for Mining Signals.

the engine-man immediately disconnected that signal, so that the others were not affected.

The great objection to this plan is, that it throws too much dependence upon the engine-man's memory. If he omits to disconnect, the other signals are thrown out of gear, as the whole strength of the battery will usually be absorbed by the one signal, and the battery itself will rapidly work down. Added to which, the fact of such a large number of cells being in use to-gether, will cause them to break down more quickly

o

under a sudden strain. The whole difficulty would be
overcome, of course, by making the cells very large
indeed, so that a connection on one circuit made no
difference, and using only the num-
ber required for the longest road.

The Insulators used on engine-
plane signals, are usually of the
small reel form shown in Fig. 82.

Fig. 82.—Reel Insulator
for Engine-plane
Signals.

A long screw passes through the
hole in the centre, securing it to
the prop or beam ; and the iron
wire is laid in the groove and securely bound in.

The usual method of fixing is, to place two of these
insulators on one prop, or beam, about from 4 in. to 8
in. apart. The number of insulators used will depend
upon the nature of the road. Going round a curve,
a pair may be required every yard, while, on a
straight road, from 8 to 12 yards may be spanned.

One simple rule may be taken as a guide. The iron
wires must touch nothing but the insulators ; and the
number of the latter should only be sufficient to
ensure the wires being strained tight.

Figs. 83, 84.—Insulators arranged to Carry two Wires.

Other forms of insulators that are sometimes preferred for engine-plane signals are shown in Figs. 83,
84, the object in using them being to save the neces-

sity of two fixings. In the Author's opinion, these insulators are not so well adapted for the work, as they require a more expensive fastener, and there must be more of them, otherwise there will be great chance of the wires getting into contact. That is to say, the spans cannot be made so long, as the wires are closer together and may more easily sag into contact.

Fig. 85 shows another form, known as Bright's, that is occasionally used on wet roads, the configuration of the insulator offering a higher resistance to the leakage current than the small reels; but the insulator itself is very much more expensive, and all that is sought for can be attained much more efficiently and cheaply in other ways.

In some cases plain galvanized iron staples have been used, with apparent success, with a piece of

Fig. 85.—Bright's Shackle Insulator.

india-rubber or brattice cloth for an insulator. There are, however, two grave objections to this. One is, the almost certain leakage from wire to wire; and the other is, the faulty mechanical construction. The office of the insulator is two-fold. It is a support as well as an insulator; and its primary object is support.

One great evil that suspended wires have to contend with everywhere is, their tendency to run back when a break occurs. This may be particularly inconvenient on an engine plane, as a passing tram may catch the wire when it lies on the ground and carry it along, pulling down a large portion of it. Where wires are properly bound into their insulators, they do not run back much, and are easily pulled up again. Where staples are used, there is really nothing to hold the wire, and it runs back to the terminating shackle.

Fig. 86, shows the shackle insulator generally used for the termini of engine-plane signals. It consists of an earthenware insulator, made very much more strongly than either of the forms already described, held between two galvanized iron straps by a galvanized iron bolt passing through the hole in the centre. A coach screw passing through holes in the straps and into the prop, holds it to the latter. The whole forms a very strong apparatus, either for commencing from or terminating at; the wire being twisted round the groove in the insulator. It is also very use-ful for junctions, giving a much neater and stronger appearance to the wires, than the cheaper but rougher plan, some-times employed, of using only reel insulators.

Fig. 86.—Single Shackle used for Terminating.

The covered wire used for connections between the bell and the battery, between the bells and the iron wires, etc., are of copper covered with gutta-percha, india-rubber, or Callender's bituminous compound. The wire should never be smaller than No. 18 gauge; No. 16 is better than No. 18. The reason is rather a

mechanical one than electrical. So far as the resistance of the copper wire goes, No. 20 would answer in
the majority of cases; but as it has frequently to
stand rough usage, No. 20 would be broken very easily.
Further, it has already been mentioned that wires
become brittle where they are connected to terminals
after being in use, and may be broken by accidents,
such as knocks, vibrations, etc., to which they are
very liable in mines. It will therefore be easily understood that for this reason also, the larger the gauge the
better, as breakages will be less frequent. There is
also another reason, of even more importance, for not
making wires too small in mining work. They are
very liable to be attacked by moisture, wherever they
are exposed, as at the terminals of bells, keys, and
batteries. At battery terminals, also, the battery
liquid is very liable to be spilt over them unnoticed,
even by the most careful men. Where the wires are
small, the chemical action of the moisture on the copper, assisted as it usually is by the current, soon parts
the wire; and therefore, unless the signal has very
constant attention, it is continually failing from this
cause. Stranded wires are even more liable than small
solid wires to this failing, as the moisture or battery
liquid attacks each wire of which the strand is composed.

For the shaft, where the engine is on the surface,
one or both of the wires may be insulated, and should
be of copper covered with gutta-percha. The Author
prefers to have one covered copper wire and one

naked iron wire in the shaft, and for the following reasons :—

Firstly, It is cheaper.

Secondly, The signal so constructed is less liable to get out of order.

The first of these propositions is self-evident. The second would almost seem to run counter to all that has gone before; but a little consideration will show that this is not so. The covered wire in the shaft is the weak point of any signal in which it is employed, because it is so liable to be damaged; both by the water in the pit, which is often impregnated with powerful metallic salts, and also by falling coal and other bodies; and so difficult to get at for repair. Moreover, of all faults, a break in a covered wire in a mine shaft is the most difficult to find. The iron wire can always be run in one, or at most two lengths; need only be stapled to the side of the shaft, out of the way, and rarely gives any trouble.

The covered wire will invariably give frequent trouble, unless it be strong, well insulated, protected from mechanical injury, and very carefully laid. No staples should be used; or, if this is unavoidable, the wire should be well wrapped before being stapled.

The best plan, according to the Author's experience, is, to have a copper wire not smaller than No. 16 s.w. gauge, well insulated with gutta-percha, the larger the gauge of the wire and the thicker the insulation the better, the latter never less than No. 6 ; and outside of this a covering of hemp or jute, as much as ever you can get on. Then place your wire in boarding grooved to receive it; the boarding being firmly secured to the side of the shaft; and either fix the wire in place by small vertical pegs of wood, or by making the groove just a fit and gently tapping the wire into it. The boarding may be covered over after if desired; but this is not absolutely necessary, as, if the wires are set well back in the grooves, the latter being made deep enough, and small boards be placed over the joints of the planks, it will not be easy for anything to get at the wires, while they are more readily inspected in case of accident than when covered.

With these precautions, however, accidents are very rare, and wires will go for a good many years without any trouble whatever.

Where the covering of shaft wires is too thin, even of such material as gutta-percha, the wet always penetrates through it in time, causing leakage; and if there happen to be a weak place in the wire, such as some trifling flaw in the drawing, that would give no trouble except in the presence of wet, it will be sure to be found out, the wire parted there inside the covering, and a failure caused that will be exceedingly difficult to discover and repair. It is for this reason also that the Author recommends a thick outer covering of jute or yarn, to prevent the percolation of moisture through to the insulating covering as much as possible.

India-rubber, whether vulcanized or not, will *not* stand as a covering for shaft wires, unless the protecting covering outside of it be absolutely impervious to moisture. India-rubber invariably softens with time, allowing the wet to penetrate, causing leakage and the other troubles.

Iron pipes are not to be recommended as a protection for covered wires in mine shafts. They are more expensive than wood, cost more to fix, the wires are often damaged in being drawn into them, and during the process of jointing, and when once fixed you cannot examine your wires without an immense amount of labour and trouble.

Never . joint a covered wire *in the shaft*, if it can possibly be avoided. Use a whole coil longer than you require, and cut the end off. Or, if you must have a joint, make it on the surface, under cover.

The joint in a covered wire should be made as follows :—clean the two ends for about 1½ inches each, make a bell-hanger's twist as already explained, and then cover either with gutta percha or Chatterton's compound, carrying the cover well back on each side of the joint, and making it one homogeneous mass with the original covering of the wire. Or, take a strip of plain india-rubber, and put it on under tension, making three or four layers, each crossing the one below, and all starting from well back on each side of the joint. Put what sailors call a whipping, with fine twine, on each end, cover over all with primed tape, and then with jute. Jointing covered wires is *most* difficult, and requires very great care. Soldering is not necessary ; but if you do solder let it be with resin.

In the Author's experience, however, unsoldered joints stand best in the copper wires used for signalling. The heat employed in soldering is apt to destroy the temper of the wire, and cause it to part readily later on. If spirits are used too, a little is almost always left, with the same result.

Naked iron wires have been used in mine shafts, but not with success. If they rest on insulators, the latter soon lose their insulation resistance, owing to the accumulation of damp dirt on their surfaces. If they are simply hung from the top and strained at the bottom, you will have some leakage at the bottom, and you will be very liable to lose your wires periodically by breakage. When the wire does break, it falls to the bottom of the shaft in a tangled coil, and is usually irrecoverable.

Causes of Failure of Engine-plane Signals.—Engine-plane Signals may fail in two ways. Their bells may refuse to ring when the wires are placed in contact, either from the hammer not coming up to the bell, or from its being already there ; and in some cases the signal will refuse to ring more than once, because the hammer does not go back after a signal, so as to be ready for a second.

Where trembler bells are used, the latter failing resolves itself into continuous ringing.

The cause of the hammer refusing to come up to the bell, is of course the current passing round the electro-magnet is too weak to generate sufficient magnetism to overcome the inertia of the armature and hammer.

With badly-constructed bells, as already explained, the cause of failure may be in the bell itself. Assuming this to be properly constructed; the weakness of the current may be due to decreased E.M.F., or to increased resistance, and is often partly due to both. The E.M.F. of the battery has fallen, owing both to the decreased initial E.M.F. and to increased internal resistance; and the external resistance has also increased. The increase of the external resistance is due generally to dirt on the wires, and to defective joints. Therefore, if a signal begins to ring feebly, examine the wires carefully, clean them if possible, and make all joints good.

Where a number of falls have occurred within a short distance, giving rise to very frequent joints, it is best to cut that piece of wire right out and replace it.

In some mines, however, it is not possible to keep the wires free from joints for long together, nor is it possible to keep them clean. In these cases, it is of importance that the wires should be as large as possible, so as to decrease the working resistance, and to decrease the breakages; they should be kept as clean and free from joints as practicable, and the rest must be made up in the battery and the bell; the latter having the powerful electro-magnet already described, and the former being composed of large-sized cells; and, where there is not much leakage, plenty of them.

The first thing to be done, then, when a signal fails, is to test the battery with a detector, as already described in the chapter on Galvanic Batteries, disconnecting and replacing any faulty cells.

Having got the battery up to its full strength, or as near that as possible, try the signal just outside the engine-house, where the iron wires are connected to the covered wire leading to the bell and battery.

If the bell now rings as loudly as at first, while before the battery was done up, it either rung weakly or not at all, probably the fault is repaired for the present.

If the bell rings well near the engine-house, and gradually becomes fainter as you recede from it; the battery is not sufficiently strong ; there are either some faulty cells in it, or the number is not large enough. If the bell rings well at a certain point, and does not ring a short distance farther off, or only rings weakly at that point, examine the wires carefully between, probably there is a bad joint somewhere, introducing such a high resistance into the circuit that the current is weakened below its working strength.

A detector galvanometer is of great service in making tests on the line wires. By taking the deflection of one or both circuits across the iron wires at the engine-house, and then at different points on the road, a sure guide is obtained. Where a sudden decrease in the deflection occurs between two points, the fault, or at any rate *a* fault, lies between them.

If the bell rings well after not being in use for some time, and gradually fails during the day, possibly ringing once, when two raps are intended, the battery is at fault. There are either some faulty cells in it that require renewing; or, if it has not been long in use, it is too weak for its work. That is to say, either the cells are too small, or there are not enough of them. If the cells are of proper size, adding a few more will usually put matters right, unless the bell is weak too; but if the cells are small, adding more will only increase the trouble, though it will apparently make the ringing better for the time. The reason has been already explained.

If the bells are weakly constructed, with small electro-magnets, there is no help for it, you must either alter them or do up the battery as often as the bell begins to fail. Adding cells will not help you in this case, whatever size they may be; though the larger they are the less attention they will require.

If the bell does not ring when you are sure that you have made the battery good, even close to the engine-house, the fault must lie either in the bell itself, in the switch for disconnecting at night, where one is used, in the covered wire or the connections. Carefully examine the whole of the covered wire, looking for the green verdigris already mentioned. Examine the connections between the covered wires and the iron wires, the connections to the bell, switch, and battery, the working faces of the switch; see that all the moving parts of the bell work freely. If no cause is apparent, test as follows, using a detector galvanometer if you have one.

Take a piece of covered wire you know to be sound, connect it to that end of the battery which is not

connected to the bell. Connect its other end,—the
ends being clean of course,—to one terminal of the
galvanometer, if you use one ; if not, use it to test. If
you use a galvanometer, connect another piece of
covered wire to its other terminal, and use the other
end of this wire to test. Now proceed as directed in
testing for faults in domestic electric bells ; and remember the rule, as before :—that when you find two
points, at one of which your bell rings, or you have
evidence of a current, while at the other you have no
evidence of current, or you have evidence that the
current strength is very considerably reduced, the
cause of failure of your apparatus lies between those
two points. Following this rule, commence at the
connection to the iron wire,—the one connected to the
wire leading from the bell,—and work back towards
the bell, until you obtain evidence of a full current.
If you find the fault in the bell itself, pursue the same
course to discover the exact locality of the break.

The most likely fault, however, in cases of this kind,
is, the covering of some portion of the wire having
been damaged, moisture has crept in and parted it.

Where there is a bell at the inbye end of the engine
plane, failure may be caused by a disconnection in that
bell, in its earth wire, or in the earth connection at the
engine-house. In this case, first connect the iron wire
leading to the inbye bell to the earth side of the
battery, and test as for a signal with only a bell in the
engine-house. If the fault is not discovered, carefully
examine the earth connection before ground is reached

and in its connection with ground ; and if there is any doubt as to its efficiency, replace it by another, or better still, duplicate it, the second one being perfectly independent of the first. If the fault is still undiscovered, seek for it at the inbye end. Leave your wire leading to the inbye bell connected to earth at the engine-house. See that your line wires are in proper order all through. Carefully examine the inbye bell and the earth connection there and duplicate the latter if it can be done easily, then proceed to test as follows:—Connect one wire from your detector galvanometer to the iron wire that is connected to the battery at the engine-house. It will easily be known, of course, by its being the one that is not connected to anything, unless there is a key at that end.

With the other galvanometer wire, cleaned as usual, make connection to the line wire with the bell in, and observe the deflection of the galvanometer, and note if the bell rings. If you have the long circuit of the galvanometer in use, the bell may not ring, owing to the additional resistance of the galvanometer wire ; and it is therefore wise to test with both circuits; the extra time taken up is not serious.

It may be mentioned also, *en passant*, that it is often useful to place the long circuit of the detector galvanometer in circuit in many tests, to roughly represent the resistance of the line wire. If the bell does not ring, test beyond the bell, and follow down to earth, noting any change in the deflection as before. If you get a deflection beyond the bell, and none or a greater decrease than the resistance of the bell coils should account for, test through the bell itself as already described.

If you get a deflection on the outbye side of the bell, but the bell not ringing, even with the short circuit of the galvanometer, and it rings when you connect to the earth side, the fault is evidently in the earth wire. It may, of course, be at either end, but is more likely to be at the inbye end.

Seeking for a faulty earth is perhaps the most difficult of all tests ; the only method being to eliminate every other possible source of failure.

Where a number of bells are connected to one engine-plane signal, whether in series or in parallel, the same plan is adopted, commencing at the engine-house, and working outwards, testing at each station until the fault is discovered.

Where the engine is on the surface, always test at the pit top first, with a detector galvanometer if possible ; or if not, by hearing the bell ring ; and then test at the pit bottom. If the bell rings, or you obtain evidence of a current at the top, and none, or greatly reduced, at the bottom, the fault is in the shaft, and will be the most troublesome fault you can have to deal with. Be sure that it is in the shaft before you test ; and not, for instance, in the bell at the pit bottom, where one is used. This may be discovered by testing on each side of the bell. If you obtain a deflection on the shaft side of the bell and none the other, the fault is in the bell.

Having traced the fault to the shaft, get as much
light as you can, go through the shaft *very* slowly, and
examine every inch of wire ; if possible, passing the
covered wire through your hand. As mine shafts are
always wet ; and the wires, especially when not covered,
become coated with the material that is washed off the
wall of the shaft and finds its way there from various
sources, it is not an easy thing to find a fault in a
covered wire. Usually such a fault is due to the covering being damaged and moisture having penetrated ;
but the insulation covering holds the wire together, and
the sludge outside often covers up the green verdigris
that is such a sure guide in other cases.

Examine the iron wire also, in case there should
have been a bad joint, or a flaw that has rusted in two.
This is usually a very easy matter. Having examined
the wires *very* carefully, pulled at all the staples if any
are used, if no sign of a break has been found, there
is nothing for it but testing at different points, *in the
shaft ;* making connection between the covered wire,
bared for the purpose, and the iron wire ; or between
the two covered wires where they are used. As the
cutting of the insulation covering will probably lead
to future failures unless very great care is taken to recover the bare place, it is always wise to spend a little
time examining the shaft wires even three or four
times, before resorting to this expedient. If it is in-
evitable ; proceed to about half-way down the shaft,
clean a place on the iron wire and connect one of the
galvanometer wires or the bare end of a piece of stiffish

wire to it, then *very* carefully shave a piece of the covering of the insulated wire off, so as just to expose the copper. The latter must be sufficiently visible and clean to make good connection, or no current will pass, and it will apparently show that the fault is above you when it is not. If, having obtained good connection between the iron wire and the covered wire (either through the galvanometer or the short piece of wire), you find no indication of a current, or only of a very feeble one, the fault will be in the upper part of the shaft. Be very careful, in making this test, not to be deceived by dirty connections, etc., or the fault may take a long time to find. If a full deflection is obtained, or the bell rings, the break is in the lower half. Proceed to test half-way between this spot and either the top or bottom as the case may be, and so on until two points are found, as before, between which the break must lie. Usually a more careful examination of this piece will now reveal the fault. If not, cut the piece out and replace it by a new one.

After these tests, be very careful to cover the wire up *immediately*, as described for covering joints. The piece that is put in also must be jointed very carefully, it being almost an impossibility in many shafts to prevent moisture being left inside the covering. After making joints of this kind, it will be wise to examine them as often as opportunity offers, and remake them whenever they show signs of giving trouble.

Leakage Faults.—The other class of faults that are

P

met with in engine-plane signals, where the hammer remains against the bell, or continues chattering if a trembler, are due to leakage or actual contact between the wires.

This class of fault is more troublesome to find than the other, as it necessitates repeatedly going to and fro, to ascertain if the fault is off, unless some signal can be arranged not involving the use of the wires. The reason for this is, that a test can only be made near the battery.

If the hammer be held firmly against the bell, the first thing to be done is, to carefully examine the wires on the engine plane and observe if they are in contact with each other, or with some other metallic body, such as a pulley or a wire rope. It may happen that one wire will be in contact with a metallic body in one part of the engine plane and the other wire is in contact with another such body in another part of the plane, and that these two connections complete the circuit; because, although the connection is not immediately apparent, there may be, and is if a short circuit is formed, a series of metallic connections between these two bodies. It is therefore wisest to remove all naked wires from contact with everything but the insulators. Coal, shale, slate, and wood are not good conductors, but they are very bad insulators, more particularly when they are saturated with moisture, or even have a film of moisture on their surfaces. Therefore do not allow the wires to touch the props, the roof, or the sides of the road. In damp mines,

too, the film of moisture which forms on the cold surface of the insulators is a conductor of high resistance,
which causes a serious leakage when multiplied many
hundred times, as it will be on a long road. Further,
at the point where the insulators are screwed to the
props, fungoid growths sometimes form over the whole
of the insulator, and their roots extend into the fibres
of the wood. Coal-dust and dirt also frequently lodge
on the insulators. Therefore, when a signal begins to
fail, besides seeing the wires clear of everything, wipe
the insulators as clean as possible; and if this be done
periodically it will save trouble. Rather a common
source of leakage is the practice of tying up the wires
with string, after insulators have been broken. String
is cheaper than an insulator, and it is perhaps accidentally discovered that the replacement of a single pair
of insulators by string attachments does not affect the
working of the signal; as it would not if there is the
proper reserve in battery and bell. The exchange is
repeated as often as an insulator breaks, until the signal begins to fail, owing to the extra drain upon the
battery caused by the leakage current. Therefore, if
a string attachment is made, let it be replaced at the
earliest opportunity by insulators; and if, when the
signal fails by the hammer coming up to the bell,
insulators are wanting anywhere, let them be immediately fixed.

Supposing that a careful examination does not lead
to the discovery of the cause, and a careful overhauling
of the wires leaves the hammer still up at the bell

when the signal is connected, or remaining there after
a signal has been given, proceed to test as follows:—

Take several convenient points—as, say, at the top of
the shaft, the bottom of the shaft; where the engine is
on the surface; at junctions, at joints in the wires; and
break the main circuit at each of these places in succession, observing the effect upon the hammer. If the
hammer falls back after a particular portion of the
signal wires are thrown out of circuit, while there is
no apparent difference when other portions are thrown
out, the fault will lie in that portion; and a further and
more careful examination of that part of the road
should be made.

To have a sure guide, it is better to use the long
circuit of the detector galvanometer, inserting it in
circuit; then, by disconnecting at successive points,
working outwards, and re-connecting, and observing
the difference in the deflection after each break, one
is enabled to see exactly where the leakage is. If, for
instance, as already explained, the deflection is reduced
to zero when one section is cut off, the leakage will
all be in that section; but if, as more frequently
happens, the deflection is reduced by the cutting off
of each section, roughly in proportion to the length of
the section, the leakage will be general throughout the
signal, and the remedy must be sought in a more or
less general reconstruction.

If the leakage be traced to the shaft wires, it will
usually be necessary to replace the covered wire
throughout; but an attempt may be made to find the

bad place, should such exist, by proceeding as in testing for disconnection,—cutting the wire in two at each point, instead of merely removing a portion of the covering. Thus, if all or nearly all the deflection disappears on cutting, say, three parts of the way down the shaft, it will be evident that the other quarter is causing the leakage.

Should this turn out to be the case, the same precautions should be taken for preserving the joints from being eaten in two by the moisture of the pit, that have been already described for disconnection faults.

It unfortunately happens, in many wet mines, that it is impossible to avoid leakage on the roads. Leakage in the shaft can always be provided against, by suitable covering and protection being given to the wires. On the roads, however, where it is essential that the wires should be left bare for the purpose of signalling, and a certain number of insulators must be used for mechanical reasons, a heavy leakage current is often unavoidable, especially on long roads, where the supports are multiplied.

In other cases it is difficult to maintain the insulation perfect, owing to frequent falls.

In such cases there is nothing to be done but provide for the leakage current in the construction of the apparatus used, keeping up the insulation as much as possible.

The first thing to be done, then, is to reduce the leakage current by reducing the number of cells. This will decrease the strain upon the battery. Next, if they are not so already, use the largest size cells obtainable, so that they may maintain their normal strength notwithstanding the drain of the leakage current, and so keep the signal up to working strength. Next, if the leakage current is still sufficiently strong to hold the hammer up to the bell when brought there by a signal, bring into play the buffer spring, or its equivalent that should exist in every Mining Signal Bell, and one form of which is shown in Fig. 31.

If the tension of this spring be increased, by means of the screw provided for the purpose, the hammer will not be allowed to remain on the bell after giving a signal, as the spring will throw it off. Another plan is, to alter the screw of the bracket in which the hammer works, shown in Fig. 31, so that the armature cannot come so near the poles as before. If this is done, either the hammer must be bent forward, or the bell dome moved towards it. It must be understood, that by either of these plans the battery has more work thrown upon it, but that it is the simplest and the easiest plan of getting over a troublesome difficulty. With trembler bells it is by no means so easy to grapple with a leakage current, without altering the construction of the bell ; and with this type of bell the fault is particularly troublesome, as it causes the bell to be continually ringing. Something may be done, on the same lines as those indicated for the single-stroke bell. The armature may be set farther from the poles, in some forms of bell, by bending the spring to which it is attached, and the hammer ; but it is hardly wise, unless it can be done with great care, and by a skilled mechanic.

The writer designed another plan some years back,

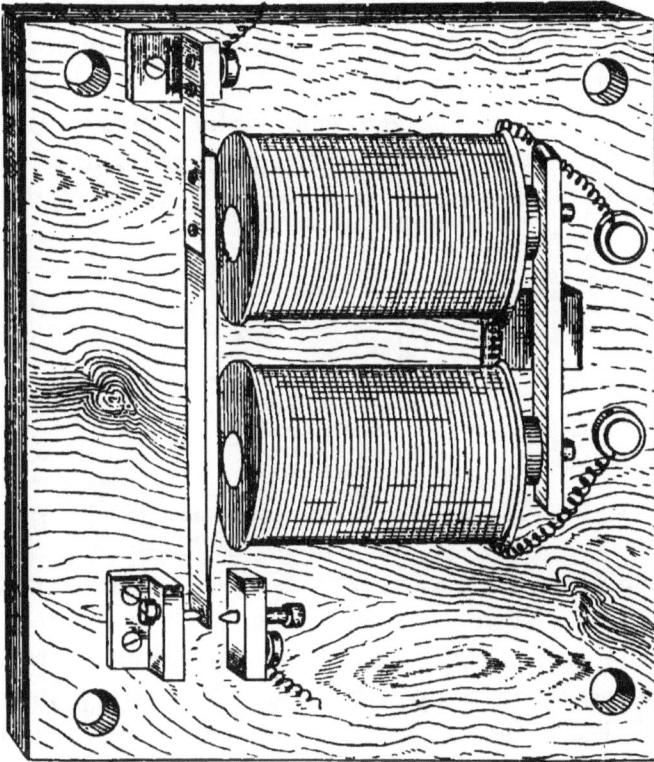

Fig. 87.—Relay for Use with Colliery Signals.

for a special case where the leakage was excessive and could not be got under, in which a Relay is used, an apparatus well known to telegraph engineers, and one that has done good service in numerous cases some-what analogous.

A Relay is shown in Fig. 87. Its construction is similar to that of a single-stroke bell, except that its armature carries no hammer, and that the whole is designed to give the working current as little to do as possible, and yet to be under control. To effect this, the armature is made very light, hangs vertically, and only carries a light platinum-tipped contact-piece at its extremity, which is faced by another platinum-

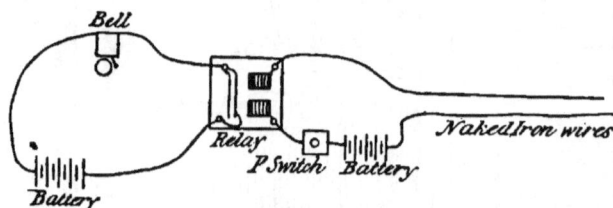

Fig. 88.—Diagram of an Engine-plane Signal, with Relay.

tipped contact-piece, as shown. The armature is con-trolled by a spring whose tension can be regulated.

Where the Relay is used, the battery cells are still made as large as possible, but the battery itself is divided. A few cells only, say four or six, according to the length of the road, are connected to the Relay and the engine-plane wires ; the rest of the battery, consisting of as many cells as you like to put on, is connected to the bell, through the Relay contact. Fig. 88 shows the double connection of Relay and Bell. The action is as follows:—When a signal is given on the road, the current from the small battery passes round the Relay coils, causing the magnet to attract its armature, and bringing the contacts together. The Relay contacts coming together close the circuit of the bell and the large battery, causing the former to ring. For practical purposes the two operations occupy no more time than the one where only the bell circuit is used. It will be seen that this plan reduces the leakage to the lowest possible limit, and yet allows of the signal being worked with a very small current. The operation of pulling the armature up to the fixed contact, a small fraction of an inch away, involves only a very small amount of mechanical work; while it is always possible to put sufficient tension on the regulating spring to prevent the armature from working under the influence of the leakage current, and yet to have sufficient margin of power for it to make firm contact when the regular working current passes through its coils.

The objections to this apparatus are, that it destroys the perfect simplicity of the engine-plane signal, and that it cannot be used for signals with two or more bells, without placing a battery and relay near each bell; which might be inconvenient in many cases, and would always give extra trouble in maintenance.

The tricks that are played with engine-plane signals are a great source of annoyance. Mischievous boys enjoy the fun of tying the wires together and stopping the work for the time. The Police Court is about the best preventive, if one of the offenders can be secured; and it usually happens that as the novelty of the thing wears off, the trouble disappears; but in obstinate cases, the following arrangement, that was worked out by some of the officials of the Rhymney Iron Co's. Collieries, and by those of the Mardy Colliery, Ferndale, will answer all purposes. The engine and hauling rope are made to form part of the circuit, taking the place of the return wire; and the two iron wires, connected together as one, form the line wire.

A switch near the engine-driver controls the arrangement, the signal being made by means of a piece of chain or wire, long enough to reach from the wires to the rope or rails, carried in the rider's pocket.

The objection to this arrangement is, that the switch and extra connections are possible sources of failure.

Underground Tapper Signals.—At several collieries in the North of England, a modification of the engine-plane signal is used, arranged to signal from station to station, and sometimes from intermediate points between two stations to one or both. A road is divided into sections, a man being placed at the junction of two sections, with two bells and keys under his charge. He receives a signal on one bell, and transmits it to the next station, when he is ready, by the key of the other signal, just as railway signal-men do.

This arrangement necessitates a battery at each station, and what are known as Tapper Bells, *i.e.* bells with keys attached.

Where it is only required to signal from station to station, one wire is sufficient, if good earth can be obtained. This may be covered and protected by boarding.

Where it is required to signal from intermediate points between stations, two wires are necessary, as with the engine-plane signal already described.

Shaft Signals.—Electric signals are also used for the shaft, to signal from the bottom to the bank, and to the engine-house, and from the bank to the bottom. In some cases, also, signals are used from the bank to the engine-house.

In most collieries where shaft signals are used, the bank bell, and the engine-house bell ring together from the key at the shaft bottom, and the bell at the bottom from the key on the bank; the second bell in the engine-house having a separate ringing key on the bank.

These signals can be arranged in different ways. With two batteries, one on the bank and one at the bottom, and two wires in the shaft, only one of which need be insulated; with one battery at the surface and three wires in the shaft, two of which should be insulated, the line wire and the battery wire leading to the bottom key; or with three wires in the shaft, two of which are insulated, but with a battery at top and bottom. In the first two arrangements, shown in Figs. 89 and 90, double-contact keys of the forms described and shown in Figs. 68 and 69, pages 179 and 180, are used, the key in the first instance sav-

ing one covered wire. In the other two cases, shown
in Figs. 91 and 92, single-contact keys are used. When

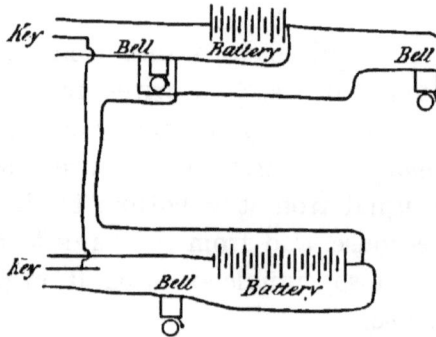

Fig. 89.—Diagram showing Connections of Shaft Signal using two
Batteries, two Wires, and Double-contact Keys.

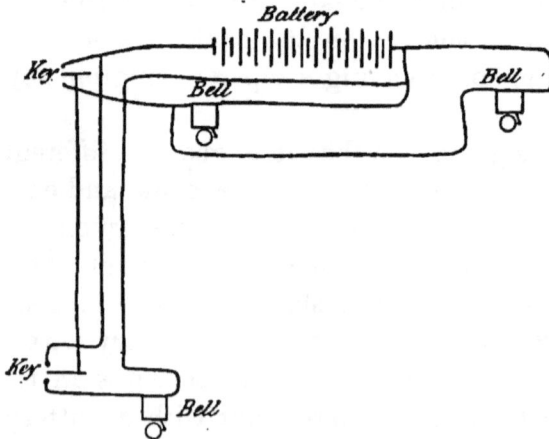

Fig. 90.—Diagram showing Connections of Shaft Signals using
one Battery, three Wires, and Double-contact Keys.

at rest, the double-contact keys are arranged to allow
an incoming current to pass to the bell at their own

station, and so to receive a signal. When put in operation, they transmit signals, the connection between the bell at their own station being broken im-

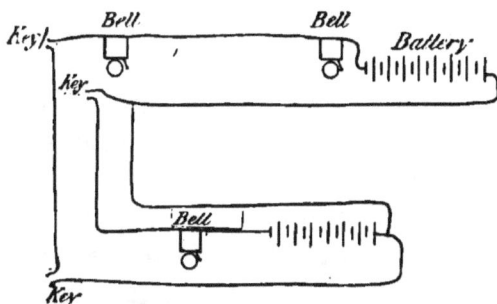

Fig. 91.—Diagram of Connections of Shaft Signals using Single-contact Keys, with three Wires and two Batteries.

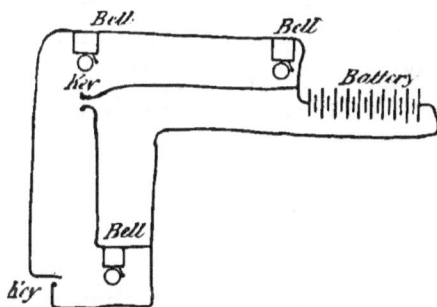

Fig. 92.—Diagram of Connections of Shaft Signals with one Battery, three Wires in Shaft, using Single-contact Keys.

mediately the lever or plunger of the key moves; while on the completion of its motion, a fresh contact is made, and another circuit closed, in which are included the battery, the ringing-key, and the bell at the other station. With the single-contact key, each key performs one office only, just as the push does with house bells; it closes the circuit in which are included the battery and the bell at the distant station.

The Author prefers the arrangement where only two wires are used in the shaft, and only one of these insulated; and his reasons are as follows :—

First, as explained in connection with engine-plane signals where the engine is on the surface, the covered wire in the shaft is the weakest portion of the whole apparatus, because it is so liable to be damaged and so difficult to repair, besides being more expensive; therefore any arrangement that reduces the number of covered wires in the shaft, reduces the first cost and the liability to failure, provided that the reduction of the number of covered wires does not necessitate the introduction of apparatus which is more troublesome. In order therefore to justify the use of only one insulated wire, it must be shown that the ringing key which renders it possible, and the division of the battery, do not give more trouble than the additional wire would do.

At first sight it would appear as though the additional covered wire in the shaft, though more expensive, would give the least trouble. It is sometimes very difficult to find a place for the battery near the pit bottom, where it can be kept dry and be easily got at. Again, the back contact of the key, unless it be very carefully and strongly made, may easily be a source of trouble. If it should not make firm contact at all times, or not go right back into its place after a signal has been given, the home bell will be cut off, so that no signals can be received.

Both these matters, however, have been successfully dealt with. Batteries are in use at even very wet pit bottoms, the precaution being taken to make the cells large, and to give them the attention necessary to enable them to stand the increased strain caused by the wet.

The difficulty of the key was, for a long time, a far more serious one, failures being frequent in the early days of electric signals from this cause; but it has been overcome by the ringing key shown in Figs. 68 and 69, page 179 and 180, assisted by the increased knowledge and the increased interest taken in the subject at mines where they are used. It will be seen that by removing the front, the working parts of the key can always be examined in a few minutes, without stopping the working of the signal. Moreover, the key can be tested, as will be explained later on, without opening it at all.

No key which does not fulfil conditions similar to the above should be used at a pit bottom. On the other hand, in addition to the danger of wires in the shaft being damaged, should there be any leakage at the battery, or any other part of the circuit, there is a strong tendency to leakage *through* the insulation the whole way down the shaft; and unless the covering be very thick, or the insulation resistance high, and remain so, the consequences may be very serious.

Any trifling fault on a battery wire is always more troublesome than on a line wire, because the circuit through the covering is always closed; and in a deep shaft the length of the wire may cause the leakage path to have a very low resistance, the resistance of the insulation covering through its substance following the law of the dimensions. Such leakage paths bring a great strain upon the battery, just as with engine-plane signals; and, in addition, a battery wire in the shaft is parted in very much less time, should the covering be damaged, than a line wire, and is more troublesome to repair. Men standing on a wet cage, handling a wet wire with wet hands, get nasty shocks, especially if their hands are cut or bruised, with the result that joints made in battery wires in a shaft, unless under exceptionally favourable circumstances, rarely last any time.

Where the battery at a pit bottom is obliged to be in a damp place, there will of course be leakage; but it will be confined to the short length of wire between the battery and the ringing key, which is by no means so serious a matter as where the wire runs the whole length of the shaft.

Leakage always occurs at a ringing key, between the terminals, when the key is obliged to be placed in the wet. At first sight the leakage path on a metal key would appear to be of lower resistance than on a wood base key; but experience shows that the wood key gives far more trouble in this respect than the metal key, the reason being probably that the films of moisture on the vulcanite insulation do not make good connection with the metal on each side, and therefore offer a high resistance. In addition to this, the film of moisture formed on vulcanite is never so large as would be the body of the moisture contained in a board.

But the principal trouble is the chemical action of the moisture upon the wires, and upon the terminals to which the wires are connected. As already explained, this will always take place in damp situations. It is very much more energetic on the battery terminal of the key than upon the others; and in many cases all that can be done is to watch the connections and clean them periodically. For this reason, whatever key is used should have all its terminals in sight. No amount of covering with gutta percha, nor leading the wire down some distance will prevent the moisture reaching the terminal; it always follows the outside of the wire and penetrates to the connection.

The bells and wires used for shaft signals are the same as for engine-plane signals; and all the remarks made with reference to the latter apply to the former, except that in some cases a 9-in. or 11-in. bell dome is used for the bank bell, in order to give a louder signal.

The bell to ring from the bank to the engine-house is usually an ordinary 4-in. trembler bell, and the key a single-contact one. Either of the keys shown in Figs. 68 to 71, will answer for this purpose.

Le Clanché batteries, Mercury Bichromate, or Sulphur Sal-ammoniac batteries may be used for Shaft Signals.

Q

For the up signal the cells should be the largest made, and plenty of them, as the up signal has usually very hard work, and the battery will run down very quickly unless both the size and number of the cells is large. The Author's practice is, to place from sixteen to twenty of the largest Le Clanché or Sulphur Sal-ammoniac cells at the pit bottom,—sixteen if it is very wet, twenty if dry,—and either ten of the largest sized cells or twelve of the second size at the surface.

Where the battery is all on the surface, it should consist of twenty of the largest sized cells.

These numbers look large, but experience proves them to be necessary if a loud signal is desired, unless the battery is to require frequent attention; moreover, it is economical in the end, as attendance forms usually the principal item in the cost of maintenance.

With bells having powerful electro-magnets, and batteries as described, the writer has known shaft signals go from six to twelve months without attention.

Iron wires alone have been used in the shaft by the Author for shaft signals, but they have not been so successful as covered wire, and for the reasons mentioned in connection with engine-plane signals, where the engine is on the surface, viz., that it is difficult with iron wires alone to provide both the necessary mechanical strength and the necessary electrical insulation resistance, for the continued efficient working of the signal.

They have a greater chance of success than with the

engine-plane signals, because the leakage path formed at each rest, from line wire to earth wire, does not affect the battery except when the signal is actually in use. It does not bring the strain upon the battery that leakage on an engine-plane signal does.

Shaft Signal Faults.—Faults in shaft signals are much the same in character as with engine-plane signals. Bells either refuse to ring, their hammers do not come up to the bell when contact is made at the key, or the hammer lies on the bell when a signal is given, and has to be put back before a second can be received, or lies there permanently as long as the bell is connected.

As before, the first is due to increased resistance somewhere in the circuit, or to decreased E.M.F. at the battery.

The latter is due to leakage or to actual contact. For resistance or disconnection,—which is, of course, only infinite resistance,—first test the battery that works the signal which has failed. That is to say, where the battery is divided, test the battery at the pit bottom for the up signal, and the one on the surface for the down signal and for the signal from the bank to the engine-house. If the battery is right, proceed as will be described; but get the battery strength up sufficiently to ring the bells well before proceeding, as otherwise it is possible to be hunting for a supposed disconnection when it is only the battery that is weak, and perhaps runs down while the testing is proceeding.

Having made sure that the battery is right, test at

the key, by making connection between the bell terminal and the battery terminal, and note if the bell rings. On the key shown in Fig. 68, p. 179, this connection can be made by passing a knife-blade across; and the spark which passes will be a rough guide as to the strength of the battery. If a detector galvanometer is available, a double test with the long and short circuit will be very useful; as, if the home bell rings through the long circuit, it will be pretty clear proof that the battery is strong enough to ring its proper signal.

If the home bell does not ring, it may be because the bell itself is faulty; but if the latter has been ringing while the other signal has failed, it will probably be the battery wire or the battery connection which is wrong. Remove the battery wire from its terminal on the key, and make connection directly, or through the detector, to the bell wire. If the bell now rings, the fault is in the key, which must be examined and repaired, pursuing the plan of bridging over each portion until the faulty one is found, as already described. If the bell does not ring when the battery wire and bell wire are connected; and it is known that the bell is in proper order as well as the wire leading to it from the key, examine the battery wire very carefully from key to battery, especially at staples if any are used, where it crosses sharp corners of iron work, etc.; looking out for the green verdigris before mentioned. If no sign of a break be visible, replace the wire, if short, by a new one and examine the old one at leisure and by daylight. If the battery be some distance from the key, test either from battery wire to earth wire with the galvanometer, or from battery wire to bell wire if no galvanometer is available, working from the key to the battery until the fault is passed, and a current is obtained. Testing from battery wire to earth wire is of course the best, as it only entails the damaging of one covered wire.

If the battery and the wires as far as the key are in order, examine the key carefully, disconnect the line wire and insert the detector galvanometer, first using the long circuit and then the short. If there are no indications of current, or they are only slight, disconnect the battery wire and connect it directly to the line wire through the galvanometer, using first the long and then the short circuit.

If there are now indications of current sufficient to ring the bells, the fault must be in the key, which must be tested through as before. Should, however, all be right as far as the key, and the galvanometer show a current of sufficient strength passing to ring the bells, yet the latter do not ring, though in proper order, the cause will be leakage on the line wire, which must be tested for as described for leakage on engine-plane wires at p. 209.

If no current passes through the galvanometer when the battery wire and line wire are connected together, leave them connected and go to the other end of the shaft,—to the top if you have been testing underground, and *vice versâ*,—and look for your current at the line wire of the key there. The line wire has now become the battery wire, and you should get a current from it to the earth wire. If you do not, either the covered line wire or the earth wire must be parted or nearly so. As before, the double test with long and short circuit of galvanometer will afford a better guide than a single test.

If the fault is in the shaft, pursue the same plan as described for the shaft wires of the engine-plane signal, carefully examining, and then testing from point to point, till the fault is passed.

If the shaft wires are right, the fault must be in the receiving key, in the bell, or in the wires leading from the key to the bell, or from the bell to earth. Examine all joints carefully, and follow the test right on from point to point, till you find the two points at one of which you get a current, and at the other you do not; when, as before, the fault lies between them.

A fault in the signal from bank to engine-house would be found, by carefully examining the key and the wires, and then testing back towards the bell, seeking a current between the battery wire leading to the key, or the line wire connected to battery, and earth.

For a disconnection, or a bell refusing to ring owing to a break in the circuit, with battery on the surface; first test the battery as before; next, if it is the up signal which is off, see that the battery wire in the shaft is perfect, by testing to earth at the pit bottom; then test each wire in succession, after having previously examined it, as with the other arrangement.

That is to say, having tested the battery wire, test the wire from the bottom key to the top bell for a break, and follow on, if it be the up signal. If it be the down signal, connect the battery and line wires at the key on the bank and test downwards. Examine the keys, and test them through before touching the wires.

Where the hammer stands up to the bell and refuses to leave it, or refuses to go back after a signal has been given; first examine the bell itself. Occasionally a slip is made in the construction of the bell, the hammer being allowed to stand on the wrong side of the vertical, when up at the bell. If this is so, it may have been put right by one of the methods mentioned in dealing with similar faults in engine-plane signals, viz.: either by increasing the tension of the spring or by setting the screw of the bracket against which the armature works a little forward. In either case, the screw may work back in time and require setting up again. A few minutes will usually put this right where two batteries are used. If the fault is not in the bell itself, it must be due to leakage in its immediate neighbourhood, and will be found best by disconnecting the battery wire at the ringing key, and working back to the battery, examining the wire carefully the whole way, and if nothing is found, cutting the wire at different points until the fault is passed. Where the wire leading from battery to key, and that from key to bell run side by side in a wet place for some distance, and this has caused the leakage, it may be reduced temporarily by placing the wires farther apart; but for permanency, both should be replaced by wires whose covering will not allow of any leakage.

If a shaft signal refuses to work, while the detector galvanometer shows that a current of sufficient strength passes out when the ringing-key is pressed, there must be leakage between the ringing-key and the bell, and the leakage path must be of sufficiently low resistance to reduce the current passing to the bell beyond it, below its working strength.

To find where the leakage is, carefully examine your wires and keys as before; and if no indications if possible leakage be found, disconnect successive sections, as described for similar faults with engine-plate signals, observing the deflection of the galvanometer after each section is cut out, until the fault is passed; the galvanometer being connected in the line wire at the ringing key. Here too, if it be a covered wire whose covering has given way to the wet, it may probably be put right temporarily by disturbing it, but it should be renewed as early as convenient, or it will cause repeated failures.

An ingenious application of electricity for signalling in the shaft has been worked out by Mr. W. Armstrong, junior, of Wingate Grange Colliery. He has had an insulated wire placed in the core of the winding rope, the ends being brought out on the cage and at the winding drum for connection. On the cage, the insulated wire is connected to a ringing key, the other terminal of which is in connection with the cage. The other end of the insulated wire is connected to a brass collar fixed to a wooden ring secured to the drum shaft. Against this collar bears a copper spring brush, much as the collecting brush of a dynamo bears on its commutator. From the copper brush, another insulated wire leads to a bell and battery; the other side

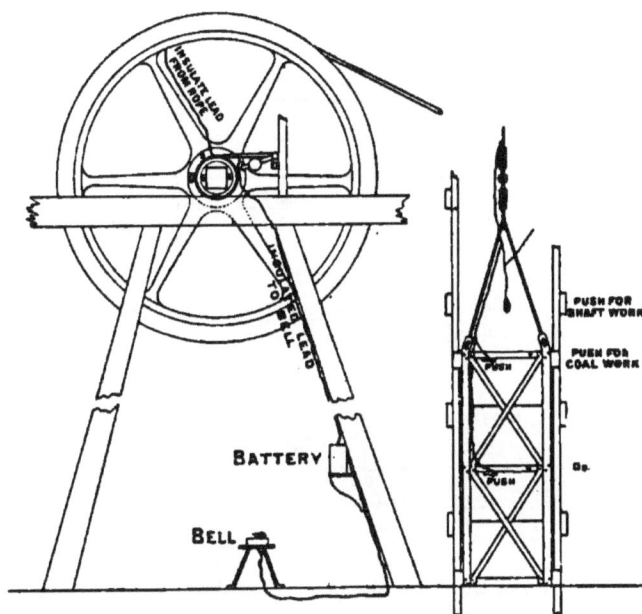

Fig. 93.—Diagram showing Connections of Armstrong's Arrangement for Signalling from the Cage.

of the battery being connected to some part of the engine framework. When the key on the cage is pressed, the circuit is completed from the battery, through the bell, the collecting arrangement on the drum shaft, the insulated wire inside the rope, the key on the cage, back through the rope itself to the engine frame, and thence to the battery. Figs. 93 and 94 show the whole arrangement in detail.

Mr. Armstrong informs the author that he has also succeeded in arranging telephonic communication between the cage and the surface, using the same

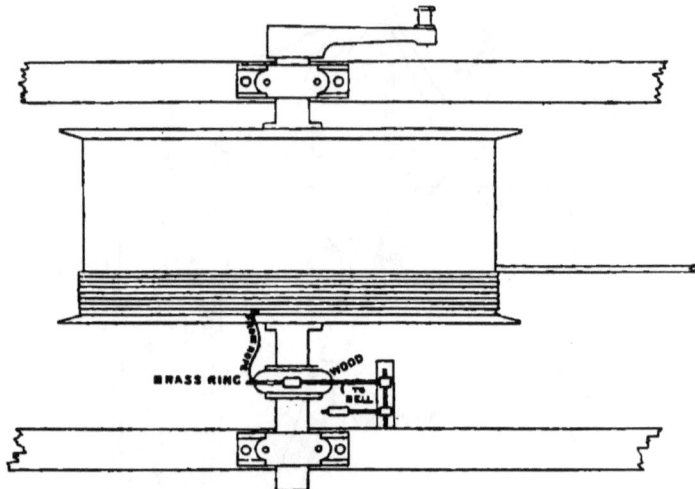

Fig. 94.—Diagram showing the Arrangement of the Connection to the End of the Wire at the Drum of Armstrong's Signal.

apparatus, but substituting telephones for the ringing key and bell.

The practical advantage of this, if successful, will be thoroughly appreciated by all who have anything to do with mines.

SIGNALS FOR CORNISH TIN, LEAD MINES, ETC.

Electric signals have as yet hardly been introduced into tin and lead mines.

As these mines are worked upon an entirely different system to coal mines, a different system of signals is required.

The arrangement recommended by the Author for Cornish mines, consists of one of the mining bells already described, 6-in. or 8-in. or larger, at the top of the shaft, connected with a battery in a convenient position, and two wires led down the shaft and connected to ringing keys at each landing-place of the form shown in Fig. 71, and described at page 182. The wires in the shaft and the connections to the keys in tin and lead mines present very serious difficulties, owing to the constant stream of water that pours down, the drippings from which are to be found in the landing-places, and often some distance into the workings.

The difficulty of the case is increased by the fact that the sinkings are often very irregular, so that it becomes an exceedingly difficult matter to fix boarding in the hard rock to follow the line of the shaft; and it is further increased by the fact that there is no regular attendant at each landing-place. The man who comes to ring the signal when the skip is wanted, has not as much light even as there is at a coal-pit bottom where electric light has not been introduced;

and he has not the opportunity of watching the formation of verdigris, as the onsetter in a coal-pit has.

There would appear to be two distinct plans upon which the shaft portion of these signals might be constructed, viz. :—by placing in the shaft two wires of stout copper, well insulated and protected from damage, leading them to the key at each landing-place, so that there would be no joints ; or, by endeavouring to dispense with insulation almost entirely, using very strong galvanized iron or copper wires in the shaft, placing them as far apart as possible, and using a battery of *very* large cells to furnish the current, and not many of them, assisting them with a relay, as already described for colliery engine-plane signals. In either case, the weak points will be the connections *to* the keys ; but if these are watched and cleaned periodically, there ought not to be much difficulty in working either plan. The writer decidedly prefers the first plan,—that with stout, well-insulated copper wires,—as it would need less attention.

Signals are working on the surface in Cornish mines and quarries, from the shaft top to the engine, and from a quarry landing to the engine. They are constructed upon exactly the same lines as those in use from colliery bank to engine, 6-in. or 8-in. mining bells being used. If the distance required to be signalled over is long, the arrangements may be as in colliery surface signals.

Surface Signals.—Signals on the surface of coal mines are used either between the colliery and a land-sale wharf; between the former and the coaling screens, where these are at a distance; and occasionally to a siding weigh-house.

They are of two kinds—the two-wire signal, and the tapper signal.

The Two-wire Signal is exactly similar to an engine-plane signal underground, except that the wires are supported on poles, to which are attached the insulator  known as the " Z," shown in Fig. 95, and which can either be screwed directly to the pole, in that case standing horizontally or slightly inclined, or pole brackets of the form shown in Fig. 96, made of galvanized iron, are nailed to the pole, the "Z" insulators being supported by them. In either case, the wires are bound to the "Z" insulators, and they are supported on poles 30 to 50 yards apart, according to the nature of the road.

Fig. 95.

Fig. 96.

Ringing keys of the form shown in Fig. 72, and described at page 183, may be used at either end, and at any particular point if desired; or the signal may be worked by pressing the wires together, as with the underground signal. Where it is only required to be operated at a few points on the road, the better plan is, to carry the wires high up on the poles, and bring connections down to the keys by means of covered wires.

Where signals are required to be made at every part of the road, the wires must be within reach for that purpose.

The bells, wire, and batteries for these signals are similar to those already described.

Faults in two-wire surface signals are dealt with exactly as those in underground two-wire signals. Examine carefully and then test from point to point. The most frequent causes of failure of these signals are, besides the regular battery faults; leakage from what telegraph engineers call weather contact, wet rags or waste stretching from wire to wire ; bad earth connections where they are used, and breaks in the covered wire, or at joints.

The Tapper Signal is used where communication is only required between two stations, such as the pit and the land-sale wharf, and not at intermediate points. A tapper bell similar to that shown at Fig. 73, and described on page 184, and a battery, are fixed at each station, and the two are connected by a naked galvanized iron wire carried on poles, and secured to " Z " insulators, the latter being supported either by pole-brackets or saddles. Terminal shackles are used at the ends of the line. Earth is used at each end ; and the connections between the line, battery, and bell are made with insulated copper wire. The earth wire may be of naked galvanized iron, carried into moist ground, with a long length buried where the ground is always moist ; or, better still, soldered to a large plate of galvanized iron. If a deep well, or a broad deep stream be at hand, either will form good earth, as also, in most cases, will the metals of the railway.

Six-inch bells are generally used for this work, and quart-sized Le Clanché or Sulphur Sal-ammoniac cells, as the work is not hard and they are not exposed to leakage. The insulated copper wire should be No. 18 or No. 16, preferably the latter, well covered with gutta-percha or india-rubber, to, say, No. 6 or No. 7.

Surface Tapper Signal Faults. — Tapper signals usually only develop one kind of fault, viz., the bells refuse to ring.

To find such a fault, first inquire carefully, as usual, whether the signal works one way and not the other, or refuses to work either way; as it frequently happens that a signal is reported as broken down when one bell fails, since communication is interrupted as much by not being able to reply to a signal as by not being able to send one.

If only one bell refuses to ring, the cause will either be in that bell itself, or in the battery or its connections at the other end. As instructed in other cases, carefully test and renew this battery. If after doing so and sending a signal, you get no reply, insert your galvanometer in the line, first the short circuit and then the long, and observe the deflection, if any. If there is no deflection on either circuit, connect the battery wire and the line wire together, as already explained in the case of shaft-signal faults, and go to the other end. You should now be able to obtain a current, and a deflection of your galvanometer at any point on the line, or at the bell, by inserting the galvanometer between the line and earth. Test thus at successive points, at the line terminal of the bell, the fulcrum of the key, the back-stop of the key, the bell coils, the earth terminal, and the connection to the earth wire. When you get a deflection through the galvanometer, you will have passed the fault. The most likely faults, other than the battery, are :—-the key not making good contact with its back stop, the connection to either the fulcrum or the back stop faulty, a break in the bell coils, the connections to the terminal screws faulty, or that to the earth wire.

The bell itself is also liable to the same structural faults that have already been described, its hammer standing up after giving a signal, if badly fitted, or its armature having worked back too far from the poles for the usual current to pull it up to the electro-magnet.

These faults are easily corrected in properly constructed bells, by adjusting the screws provided for the purpose.

Where neither bell will ring, so that signals cannot be sent in either direction, the fault is more probably in the wire part of the circuit than in the batteries, unless these have been left for some time without attention. Examine and renew the batteries first, as before. At either station insert the galvanometer in the line, and observe the deflection. If there is a break in the bell circuit, there will be no deflection on either circuit of the galvanometer. Then test each station locally. Try if the home bell will ring when battery and line are connected, which may be done in the tapper bell by inserting the blade of a knife between the lever and the battery contact. If the bell rings, it will show that the local circuit, comprising the battery, the bell coils, the key, and the battery wires, is in order. Take a wire either from the battery terminal or the battery at that end, and work outwards, making connection to the line terminal of the bell, and on to the connection to the galvanized iron wire. If the bell rings when the last connection is made, carefully examine the earth connection and then proceed to the other end, test that locally, and examine the earth wire. If the bell does not ring, take a wire from the battery terminal and work through the bell with a galvanometer in circuit, till the fault is passed. If it should be the connection to the lever from the line terminal that is at fault, it will account for both bells not ringing, this connection being included in the line circuit.

Then carefully examine the line wire, especially at joints. If nothing is found, connect the battery wire, and line wire together at one end, and then proceed to test for a current between line and earth, along the route and through the other station, till the fault is passed. In some cases it is difficult to get earth on the route ; if therefore a very careful test at each end reveals nothing, make a temporary return by connecting to the wire rope, if there be one in use and it is lying on the track—the rails, though this is sometimes uncertain, as the fish-plate joints are not good electrical joints; or, failing everything else, run out a wire temporarily for the purpose—a piece of naked No. 16 copper, or of galvanized strand, or any conductor will do that is handy and can easily be run out.

With a return wire you have complete command of the circuit, and it should only be a question of the time necessary to walk through it, and to make careful tests at various points, to find the fault with unerring certainty. .

Telephone Connections to Mining Signals.—Telephone apparatus will be dealt with in the next chapter. Any of the apparatus that will be described there can be used in connection with Mining Signals, provided that proper arrangements are made.

With shaft signals or surface tapper signals, all that is required is a connection either between the line wire and earth through the telephone switch, so that the telephone can be brought into use by merely taking the receiver off the hook, and the bell signal can be used at the same time. Or the telephone switch can be inserted in the line circuit, if the resistance of the bell is so low that it affects the speech. The former is the better arrangement. For engine-plane signals the connections are not quite so simple, nor is the working so easy. The best arrangement consists of a fixed set of apparatus near the engine, with the telephone switch inserted in the circuit, and a portable apparatus, to be connected to the engine-plane wires by the user, wherever he may be. Only experimental apparatus have so far been used in this way.

CHAPTER VII.

In order to understand how speech is transmitted by electricity, it may be wise to consider of what it consists, and the several conversions that take place in the course of transmission.

In the electric telephone, speech itself is not transmitted at all; but the acoustic energy representing it is transformed successively into magnetic and electric energy, and re-transformed into magnetic and acoustic energy.

Speech is simply an arrangement of certain sounds, or more properly certain notes, formed by the passage of the air from the lungs through the throat and mouth. Two delicate membranes, situated in the upper part of the larynx or throat, called vocal chords, vibrate as the air passes over them; and the resulting sound which is emitted, whether it be a musical note, a spoken syllable, a cough, a laugh, or the bark of a dog, is due to the formation of the mouth, throat, etc., and to the modifications involuntarily made. Thus, to lisp, or form *th*, we place the tongue between the teeth; to make a proper *s*, as in "yes," we carry the tongue to the roof of the mouth; to whistle, we contract the lips into the form of a tube; to laugh, we open the mouth wide; and so on.

The result is, that certain motions, known as sound-waves, are communicated to the surrounding air. Each simple musical note has its own sound-waves. That is to say, in order to emit any particular note, the instrument operated on must produce a certain rate of vibration in the surrounding air, causing a wave of a certain length. The sounds of the human voice, and even those emitted by animals, are merely certain arrangements of certain notes, or certain sound-waves —each wave producing its own distinct impression upon the surrounding air.

In order, therefore, to be able to transmit speech accurately and efficiently, we must have an apparatus that will accept each sound-wave that takes part in the words spoken, and that will transmit either the sound-waves themselves or their equivalent energy in such a form that they can be reproduced elsewhere in the same order and under the same arrangement as they were spoken. This is successfully accomplished for short distances, by the mechanical or acoustic telephone. With this apparatus, the sound-waves themselves are actually transmitted through a thin steel wire. The mechanical telephone consists of a diaphragm of wood, or in some cases of metal, held in a sort of sounding box. To the centre of the diaphragm is attached the connecting wire, which must be strained very tight; and, if possible, should touch nothing. Where it does touch, it is held by some form of acoustic non-conductor such as india-rubber, and is not allowed to make sharp bends where it can possibly be avoided.

The transmission of speech or of any sound in this case depends on the fact that the velocity of propagation of a sound-wave through a thin, straight, tight wire is greater than through the surrounding atmosphere, or whatever it may be in contact with, such as the india-rubber supports, etc.

Telephone Receivers.—In Electric Telephone Receivers, which are also transmitters,—as the Bell, the Gower, the "English Mechanic," the Hickley, etc.,— the sound-waves impinge on a circular diaphragm which is held in front the pole of a magnet, and which is capable of vibrating in unison with each wave. The diaphragm may be of membrane stretched like a drum, as in the "English Mechanic" and Thompson receivers, or a thin plate of mica, with a piece of iron secured in the centre ; or it may be a circular piece of thin iron, very soft and flexible. In either case, the diaphragm is held rigidly all round its circumference ; and its centre vibrates a small fraction of an inch in front of the magnet pole, in unison with the air-waves.

On the magnet pole,—which may either belong to a permanent magnet or to an electro-magnet, provided that some attracting power is present between it and the diaphragm when speech is being transmitted, —is a bobbin of fine wire, usually silk-covered copper wire of small gauge.

In the Gower-Bell Telephone Receiver there are two
such bobbins, two poles of a powerful horse-shoe permanent magnet being used.

The effect of the vibration of the iron plate upon the
wire on the bobbin is, that a current of electricity, or
more properly speaking, an E.M.F., is induced within
it; this gives rise to a current, or to a variation of
an already existing current, in whatever circuit it is
included.   The current so generated passes through
the line wire connecting the two receivers,—supposing
two to be connected together for the purpose of communicating,—through the wire of the bobbin on the
pole of the other receiver, varies the attraction of the
magnet pole for its diaphragm, and reproduces in the
latter exactly the same vibrations that were impinged
on the diaphragm of the transmitting telephone,
though in a weaker form, and thus reproduces the
exact words and tones that were originally made use
of.

Where an electro-magnet is used, the plan at present
adopted is to provide that a current of electricity shall
be passing through the wire on the bobbin, producing the necessary attraction of the diaphragm, just as
the permanent magnetism of the steel magnet does.
The impressed E.M.F. generated by the motion of the
diaphragm varies the strength of this current, the
variation, of course, being common to the whole circuit
in which the wire on the bobbin of the receiving instrument is included.

A better plan, that will probably be adopted when

the telephone patents expire, and that will enable louder speech to be attained, is, to have a separate magnetizing coil, supplied by a local battery, for the attracting force; and an outer wire for the varying E.M.F.

The "English Mechanic" receiver, the earliest form of Prof. Bell's instrument, so called because the details were published in the "English Mechanic" of August, 1879, has a membrane diaphragm and an electro-

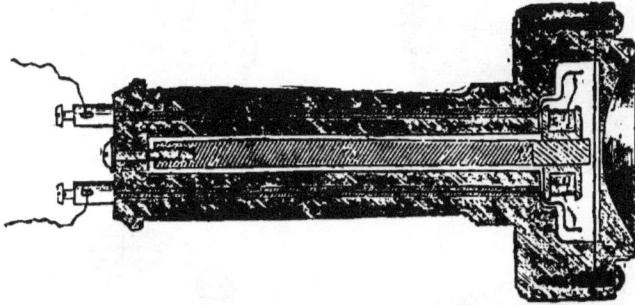

Fig. 97.—Bell's Telephone Receiver.

magnet, and is usually contained in a small wooden case, the front being made to unscrew, for the purpose of examining the diaphragm. A silk rope, similar to that already described for use in bedrooms of private houses, etc., connects it to the instrument, two of the strands of the rope containing flexible wires.

Fig. 97 shows the Bell Magneto Receiver, the one now in general use. It has a permanent magnet, a diaphragm of ferrotype iron, and is usually contained in an ebonite case, the front screwing on as with the "English Mechanic."

Figs. 98 and 99 show the Gower Telephone Receiver, which is used in the combination now known as the

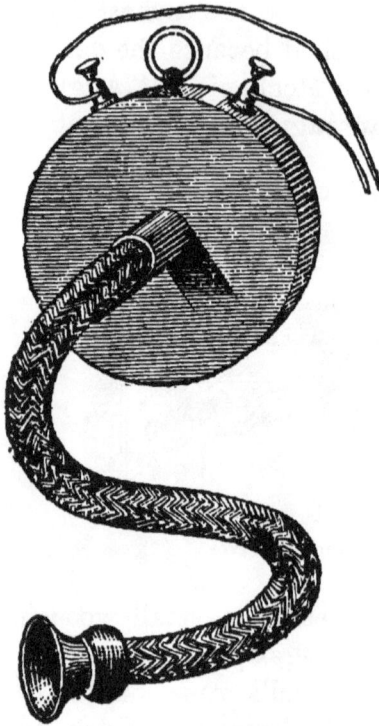

Fig. 98.— Gower-Bell Telephone Receiver. (Outside view.)

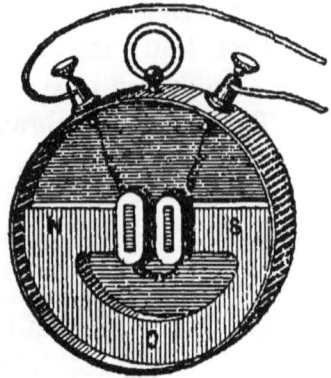

Fig. 99.—Gower-Bell Telephone Receiver. (Inside view.)

Gower-Bell Loud-Speaking Telephone. It has a horse-shoe magnet of the form shown, so arranged that the poles are brought close together. It also has a much larger diaphragm, usually of tinned iron, held in a brass collar, and one or two flexible speaking-tubes fixed so as to receive the sound-waves from the diaphragm or to carry them to it. Regulating screws are provided for altering the distance between the magnet poles and the diaphragm.

Fig. 100, shows the Hickley Receiver, which was very much used prior to its being declared an infringement of the Bell Patent. It has a diaphragm of ferrotype iron, and a permanent magnet of the form shown; consisting of one centre pole which would be either N. or S., and three others of the opposite name, S. or N., joined to an iron ring, against which the iron diaphragm was screwed. It was fixed in a wooden case, the front being secured by three screws and the cir-

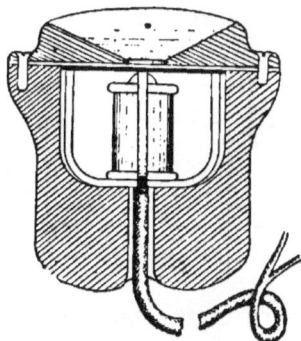

Fig. 100.

cumference of the diaphragm held rigidly between it and the case. It will be seen that by this arrangement, and that employed in the Gower Telephone, the resistance of the magnetic circuit was considerably reduced, and that will probably account for the better speech that each gave.

Figs. 101 and 102 show the instrument known as Husband's Receiver, the invention of Mr. Cox Walker, of Darlington, of which only a few were made, and

those very uncertain in their action. The peculiarity
of this instrument was, that the armature was not the
diaphragm. The former was in the form of an arch,
its opposite sides being faced by opposite poles of an
electro-magnet. On the crown of the arch rested a
disc of mica held all round its edges. It will be
evident that any motion communicated to the armature, from the poles of the electro-magnet, must raise
or lower the crown, and with it the mica disc, reproducing the sound-waves and speech. The author
has heard remarkably loud, clear speech from this

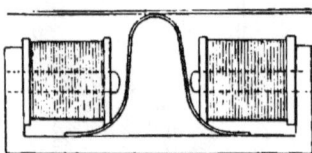   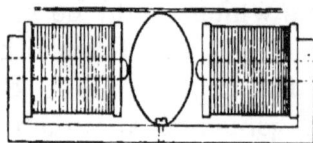

Fig. 101.   Fig. 102.

instrument; but apparently its mica discs could not
always be relied on to follow the motions of the armature. As the whole was enclosed in a wood case, the
mica disc being held in position by being pinched
between the front collar and the body of the instrument, it is probable that changes of temperature had
a great deal to do with this uncertain behaviour.

Fig. 103 shows Mr. Cox Walker's Receiver, which
he termed an electro-phone. It has an iron diaphragm
and a very small double-pole electro-magnet, with an
adjusting screw at the back to regulate the distance
of the magnet from the diaphragm. The principal fea-

ture of this instrument was, the smallness of the iron core of the electro-magnet, Mr. Cox Walker being of opinion that rapid changes of magnetism, such as rule in telephones, are more readily accepted where the mass of the iron is small.

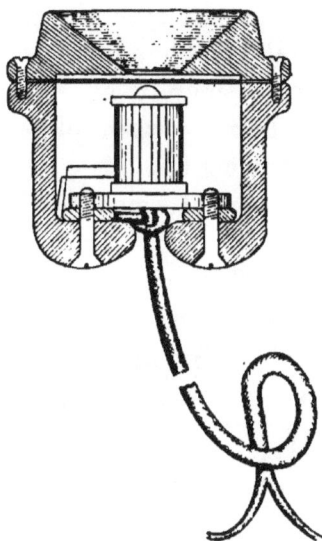

Fig. 103.—Cox
Walker's Receiver.

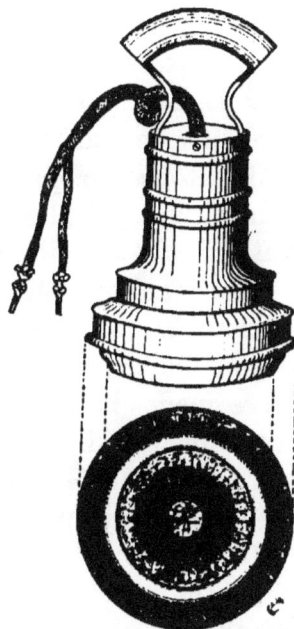

Fig. 104.—Silvanus
Thompson's Receiver.

The instrument was enclosed in a wooden case, the diaphragm being held between the front collar and the body of the case as before.

Fig. 104 shows Prof. Silvanus Thompson's modification of the "English Mechanic" Receiver. An electro-magnet is used, the bottom being surrounded by a collar of iron connected with the core. The diaphragm is a membrane stretched upon a ring of brass, and supporting a disc of tinned iron in the centre. It will be seen that by this arrangement also, the resistance of the magnetic circuit is considerably reduced ; and the instrument speaks remarkably well, though the membrane is affected by changes of temperature.

It is held in a hard wood or vulcanite case, the front screwing on and off as in the Bell Telephone.

Swinton's Telephone Receiver is merely an "English Mechanic," the diaphragm being a thin plate of vulcanite with a disc of iron in the centre.

The Stanhope and Gillett Receivers are modifications of the "English Mechanic."

The Globe Induction Receiver, of which only a few were made, is for use with microphone transmitters working with induction coils. Only Silvanus Thompson's and the Globe Receivers can be used with instruments having induction coils.

It has three bobbins, the centre one rather larger than the others, and they are faced by a membrane diaphragm stretched drum-fashion as before, and holding a strip of tinned iron opposite the three bobbins. It is held in a hard wood case, the front unscrewing as usual.

Another novel form of Receiver is the invention of Mr. Kotyra. It has no plate armature, neither membrane nor iron disc ; but it has a brass or wood diaphragm, preferably the former, for the purpose of communicating to the air the vibrations received by the electro-magnet, and forming the air-waves that produce speech, just as the ordinary telephone disc armature does.

The apparatus is constructed with a permanent magnet, with the poles bent so as to face each other, but allowing a space between for the two bobbins.

On the near side of each magnetic pole is a small bobbin of wire, similar to, but smaller than those employed in Bell's Magneto-Telephones, fitted upon a small iron core attached to the magnet-pole. One pole of the magnet is fixed rigidly to the back of the case, the other being free to move, and the bobbins are wound so that any current passing through the telephone circuit produces either mutual attraction or mutual repulsion; that is to say, currents in one direction increase the attraction of the poles for each other, and currents in the opposite direction lessen their attraction for each other, the result being that the free pole moves slightly towards or from the fixed pole, carrying the brass or wood disc with it. The periphery of the latter is held in the usual way by means of the open collar or mouth-piece through which one hears or speaks. The disc is secured to the free pole of the magnet by a screw passing through a hole in its centre.

This apparatus speaks remarkably well, and appears to the writer capable of extension, and of producing louder speech than either the Bell, the Gower-Bell, or the Hickley.

It will be noted that in this instrument also, the resistance of the magnetic circuit is reduced very considerably by the form of the magnet.

Fig. 105 shows Ader's Telephone Receiver, which was also a very successful instrument.

Transmitters.—Though all of the instruments above described will act either as Transmitter or Receiver, they are now very rarely used in the former capacity, the Microphone Transmitter being so much superior in every respect. In fact, it is not too much to say that the discovery of the microphone rendered practical telephony possible; the instruments in use previously being little better than scientific toys.

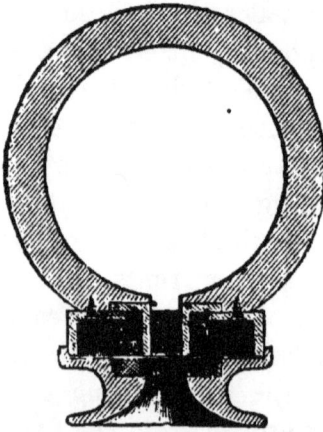

Speech transmitted by means of telephone receivers alone was so weak,

Fig. 105.—Ader's Telephone Receiver.

and so liable to be interrupted,—though no doubt a portion of this was due to want of experience with the instrument,—especially where the line wire passed near other wires carrying telegraphic messages,—that the instruments were useless for practical purposes.

Microphones.—The Microphone itself, the discovery of Prof. David Hughes, is simply a loose joint in material that will not oxidize; and it is what all electricians learn at a very early stage of their career to avoid in every other electrical apparatus. Carbon is the material now universally employed, a number of carbon blocks being drilled and screwed to a plain deal board and a number of carbon pencils fitted loosely in the holes drilled for them. Figs. 106–108 show forms of Microphones in practical use, Fig. 106 being the Gower ; Fig. 107, the Crossley ; Fig. 108, the Ader.

Fig. 109 shows the Blake, or Edison, the one adopted by the United Telephone Company and its subsidiary companies. It is very different from the others, though the principle of its action is the same. In this instrument, an iron diaphragm which faces a mouth-piece, supports at

¼ real size.

Fig. 106.—Gower's Microphone, and Carbon Pencil.

its centre a small carbon disc ; and, pressing against this by an adjustable spring is a carbon or platinum button.

Prof. Silvanus Thompson's valve Microphone differs again from the rest in the fact that no diaphragm is interposed, as in the others, between the air-waves and the Microphone, unless the tube itself be considered a diaphragm. The latter consists of three carbon pencils

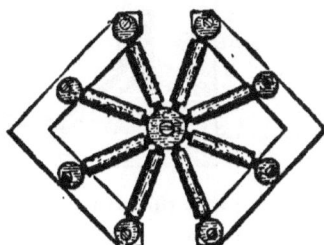

at the upper extremity of a bent tube; the carbon pencils support a small carbon ball, which dances, so to speak, upon the air-waves.

Swinton's Microphone has also no diaphragm. The carbon pencils forming the microphone are held in a

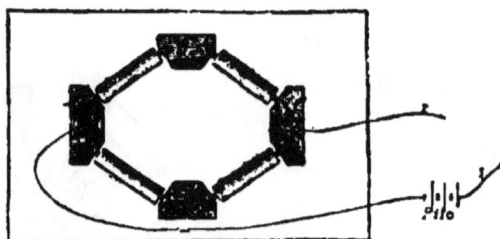

Fig. 107.—Crossley's Microphone, with Battery.

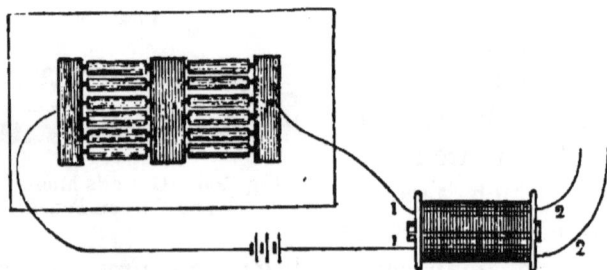

Fig. 108.— Ader's Microphone, with Battery and Induction Coil.

metal frame, suspended from the base-board of the apparatus by india-rubber.

Fig. 110 shows a section of the Hunning Microphone, which differs again in construction from any of those previously described; it having a small army of microphones consisting of a number of grains of carbon or coke held between two platinum plates as shown, the whole being enclosed, for convenience in speaking, in a wooden case similar to a telephone receiver case.

Fig. 109.—Blake Transmitter.

Moseley's Microphone is similar to Hunning's, except that small carbon plates are used instead of platinum. Several modifications of Hunning's transmitter were introduced by the London and Globe Telephone Company; but they have not been used, so far as the author is aware, since the decease of that company. They were principally designed for the purpose of avoiding the interposition of a diaphragm or separator, between the air-waves and the Microphone. Another Microphone was also introduced by the same company, in which the metal osmium replaced carbon.

The principle upon which the Microphone acts is as follows:— The diaphragm, where one is used,—or the Microphone itself, where no diaphragm is interposed between it and the air-waves,—accepts the

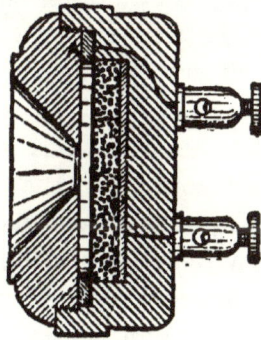

Fig. 110.

S

vibrations or undulations of the air-waves, vibrating in unison with them, and so varies the electrical resistance of any circuit of which it may form a part, in accordance with the well-known properties of a loose joint; the resistance offered by it being smaller when its parts are closely in contact than when the reverse is the case.

The Microphone has the wonderful property of responding to all the different rates of vibration of the different air-waves that go to make up speech, and of faithfully reproducing them as variations or undulations of an electric current, ready to be shown by any apparatus that is capable of being actuated by them.

All the telephone receivers described on pp. 248 to 251 are capable of being so actuated; so that, given a microphone, a battery, a telephone receiver, and some wire for connecting the apparatus together, and we have all the apparatus necessary for transmitting speech and other sounds.

It will be seen also that the interposition of the battery, by the aid of the Microphone, enables the volume of received speech to be made very much louder, as the variations of current and the variations of the attraction of the magnet or electro-magnet for its diaphragm, will be variations of a larger initial force. In fact, with good microphone transmitters, it is no uncommon thing for the received speech to be, apparently at least, louder than the spoken message.

The Microphone Transmitter.—But a microphone, a telephone, battery, and connecting wire are, after all, not the whole of the apparatus required for the efficient transmission of speech. Some method of calling attention is required; and the most natural plan is to use an electric bell. There must also be some arrangement by which the bell can be operated from the other end. Very early in the days of electric telephony, a battery for calling was added to the other apparatus; a push somewhat similar to those used in domestic bells was attached to some part of the framework or supports of the apparatus, and a bell was placed in a convenient position, at each end; so that the attendant at either end of a telephone line could call the attention of the other end, by pressing his own push, thereby making connection between the battery at his station and the telephone line wire, and causing the bell in connection with the line wire at the other station to ring.

The other station would reply, of course, by pressing its push and ringing the bell of the first station, so that the caller might know when his correspondent was at the instrument.

But now another requirement arose. The bell must be disconnected while speech was being transmitted, if a battery was in the circuit, or the ringing of the bell caused a clatter in the telephone that destroyed all chance of hearing a spoken message, every make and break in the bell-circuit causing the diaphragm of the telephone receiver to be sharply attracted and as sharply released. Thus arose a necessity for switching the bell off when the telephone was in use; and it was arranged to do this automatically by hanging the telephone when not in use on a hook which formed part of a lever. The end of this lever usually worked between two contacts, one of which connected the telephone line to the bell wire, and the other connected it to the telephone receiver. The weight of the telephone was opposed by a spring, so arranged that immediately it was removed from one end of the lever, the other end was pulled down to its contact; the spring, of course, not being strong enough to overbalance the weight of the telephone when the latter was on the hook.

Soon another requirement arose in one class of instruments. It was found that the articulation of the received speech was much clearer, the words came sharper and more distinct, when a small induction coil was used in connection with the microphone.

The plan adopted, which is now almost universal, is as follows:—One circuit is formed, including the microphone, a battery of low power, the primary wire of an induction coil, and a switch. A second circuit is formed, consisting of the telephone receivers at each end, the secondary wires of the induction coils at each end, the telephone line wire, the hook switch at each end, and the earth or return wire, whichever is used.

To save the trouble of referring to text-books for a description of the Induction Coil, it had better be given here. It will come up again when the question of the distribution of electric light by what are known as secondary generators, or transformers, is under discussion.

An Induction Coil is an apparatus for transforming a current of low tension into one of higher tension, or *vice versâ*, the conversion being accompanied by a change in the strength of the current. It consists of an electro-magnet having two insulated wires wound on it, one usually of larger gauge than the other. A few turns of the larger wire are generally wound next the iron core, and outside that, but carefully insulated from it, a much longer length of the finer wire. An arrangement is usually added for breaking the circuit of the larger or primary wire, in which a battery is included; the result being that, by the property of electric induction, as described on pages 41 and 42, each time the primary circuit is broken, and each time it is completed, a current rises in the secondary, but of higher E.M.F. between its terminals, approximately in proportion to the ratio between the numbers of turns in the two coils. The two currents generated in the secondary coil are in opposite directions, according as the primary circuit is made or broken.

It will be obvious that a *break* of the primary circuit is not necessary for the generation of a current in the secondary. A variation in the primary will give rise to a corresponding E.M.F. in the secondary; and this is the plan adopted in the telephone transmitter. The microphone vibrates, varying the resistance of the primary circuit in unison with the undulations of the sound waves, and this variation gives rise to currents of very much higher E.M.F. in the secondary coil, than are passing in the primary, so that a high E.M.F. is obtained for the line and telephone circuit, without the necessity of employing a number of cells.

It may be observed also, that the use of the Induction Coil enables a given Telephone Transmitter to be adopted for transmitting speech to any distance within certain limits, and that this may be done either by altering the battery power employed in the microphone circuit; using a larger or smaller number of cells, and thereby generating a higher or lower E.M.F. in both primary and secondary coils; or by altering the windings of the coils and using a standard number of cells, the E.M.F. being varied by varying the proportion between the convolutions of the two coils. The latter is the method employed by the G.P.O. electrical engineers and by the United Telephone Company.

With Transmitters using Induction Coils, the hook switch, as before, connects the telephone line wire to the bell, or to the telephone, according as it is on or off the hook; but it is now required to perform another operation, namely to close the microphone circuit when the telephone is in use, and to break it when the latter is not in use. Various methods of accomplishing this have been devised. So far as the author is aware, Mr. L. J. Crossley of Halifax was the first who produced a practical Telephone Transmitter, and many of his instruments are still in use, doing good service.

The shape of his microphone has already been shown in Fig. 107, p. 256. Fig. 111 shows the appearance of his Transmitter externally. The internal arrangement is as follows. A substantial hook switch, pivoted near its centre, works between two contacts, one being connected to the bell terminal on the outside of the case and the other to the small bar terminal which takes one wire of the telephone cord; the other bar

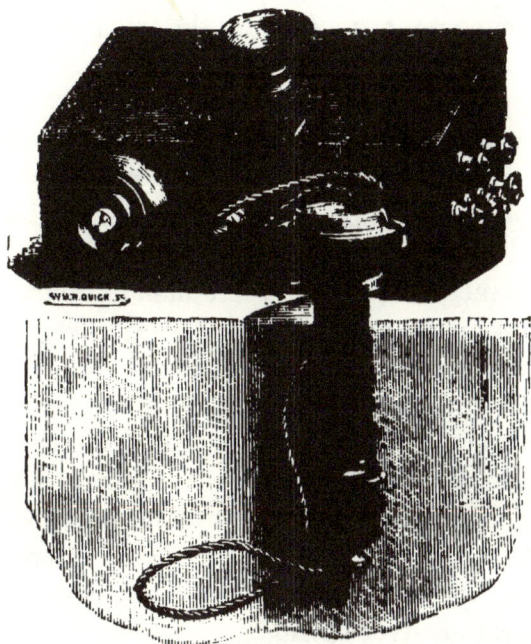

Fig. 111.—Crossley's Telephone Transmitter, with Bell Receiver.
Outside View.

terminal, to which is connected the other wire of the telephone cord, is connected to one end of the secondary wire of the induction coil, the other end of the latter being connected to the earth terminal of the instrument.

The hook switch also carries, inside the case of the instrument, an insulated bar of German silver, against which rub two German-silver springs, the whole being arranged so that when the telephone is on the hook, these springs rest on vulcanite; when the telephone is taken off the hook, the switch bar comes down. and the microphone circuit is closed by the two springs rubbing on the German-silver bar.

The top of the instrument is secured to the body of the case by screws, and the terminal screws of the microphone are connected to their proper place in the circuit by small flexible coils of wire, similar to those described in connection with bells, indicators, etc.

The connections of the microphone circuit in Crossley's Transmitter are as follows:—

One terminal of the microphone is connected to one end of the primary of the induction coil, the other end of the primary being connected to one of the German-silver springs before referred to. The other terminal of the microphone is connected to the earth terminal of the instrument, and a wire from the battery, through a terminal provided for it on the instrument, to the other German-silver spring.

A stout brass spring key at the side of the instrument, or a push in front, as shown in the figure, serves for calling and also for receiving a call on the bell, the spring itself being connected to the line, by way of the back stop of the hook switch and the bracket against which it presses, to the bell terminal. A wire from the battery leads to a terminal on the

outside of the instrument; and from there a small
wire is taken to the front stop of the key. When the
key or push is pressed, connection between the line
and the home bell is broken, and a current is sent to

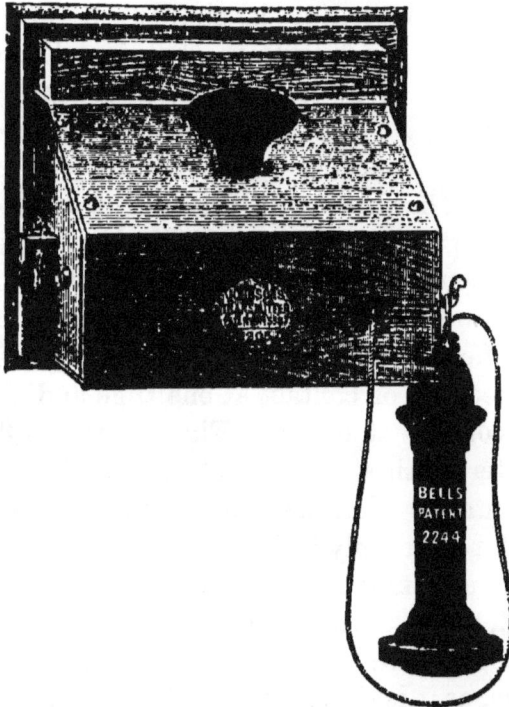

Fig. 112.—Johnson's Microphone Transmitter, with Bell Telephone.
Outside View.

line, to ring the bell at the other end ; when at rest,
with the telephone on the hook, the bell is connected to
line ready to receive a call.

Two telephone receivers can be connected to the two

bar terminals, one wire of each cord being connected
to each terminal, the current dividing between them.
This arrangement greatly facilitates the receipt of a
message, as external sounds are excluded much better
when both ears are engaged.

Johnson's Transmitter is similar in principle to
Crossley's, with the exception that Mr. Johnson con-
ceived the idea of providing a by-path for the microphone current, in case it should be too strong and
cause excessive sparking at the microphone contacts.
This by-path, or shunt coil, consists of a coil of wire
placed inside the instrument. The hook switch performs the double office of making and breaking both
telephone and microphone circuits, but it is done in a
slightly different manner to Crossley's, certain springs
being raised out of contact at one time and allowed to
fall into contact at others. Fig. 112 shows Johnson's
Transmitter, with a Bell Telephone Receiver attached.
A large number of these instruments are in use, and
they work remarkably well; but in the author's opinion
there is no advantage in the shunt coil. If a by-path
is wanted, another carbon contact is better.

The Gower-Bell.—The Gower-Bell Loud Speaking
Telephone, as it is called, is a combination of the Gower
microphone transmitter and the Gower-Bell telephone
receiver already described. It is undoubtedly the
best instrument in the market at the present time.
Fig. 113 shows its appearance externally. A mahogany case has a movable frame at the top, supporting
a thin deal board, under which is the microphone whose

form was shown in Fig. 106, p. 255. It will be seen from what has been said on electric circuits, in Chapter II. and the division of electric circuits into branches, that the star form of the Gower microphone, the oblong form of the Crossley, and the gridiron form of the Ader, are all methods of accomplishing the same object that Mr. Johnson had in view in designing his apparatus with the shunt coil. These have also this decided advantage, that each of the branches acts as a microphone, so that if any of the carbon pencils should be broken, the transmitter does not cease to operate so long as one of the other branches is intact.

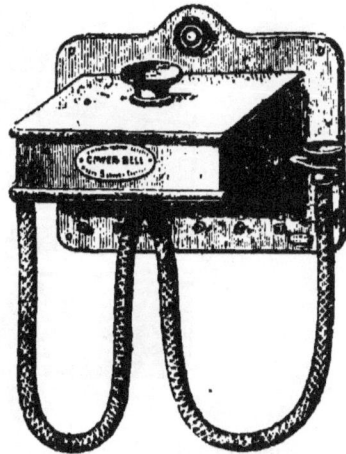

Fig. 113.—Outside View of Gower-Bell Combination.

Fig. 114 shows the internal arrangement. The Gower-Bell Receiver is placed on the base of the instrument, diaphragm downwards; the latter being held in a brass ring screwed to the base-board. Facing the diaphragm is the usual opening for the sound waves, but in this case it is filled by the open end of a Y piece of speaking tube—flexible tubes with the usual mouth-pieces being attached to each leg of the Y. At each end of the base-board are hook switches, the hook being arranged horizontally instead of vertically, as in the Crossley and Johnson instruments; the object being to provide for the support of the ends of the flexible tubes.

One of the hook switches controls the telephone and bell circuit, connecting the bell to line when the tube is on the hook, and the telephone to line, through the induction coil, when the tube is off the hook; the

Fig. 114.—Inside View of Gower-Bell Combination Instrument.
(Transmitter and Receiver.)

other controls the microphone circuit, merely making and breaking contact when the tube is off or on the hook.

An induction coil, as in the other forms of instrument, is attached to the base-board, and a trembler bell is attached to the back-board; a space being cut in the base-board to allow the bell dome and hammer to protrude.

The two hooks, one for each switch, form a very advantageous arrangement, considerably lessening the possibility of a contact failing.

A push at the top of the back-board, working the usual double-contact ringing key, completes the instrument. The spring, which is attached to line, presses against a plate of brass let into the back of the instrument, and faces a small bracket forming the battery contact, so that the whole of the ringing key is hidden and cannot be examined without taking the instrument down from the position it occupies when in use, screwed against the wall as shown. The connections between different portions of the instrument are also at the back, and therefore hidden, which is a disadvantage in the writer's opinion; though it is fair to say that neither the ringing key nor the connecting wires at the back of the instrument often give trouble, and that the arrangement gives the apparatus a very neat and finished appearance.

The Blake Transmitter, which is the one employed by the United Telephone Company and its dependencies, differs from the others that have been described in being made in two parts, both of which are placed vertically, usually screwed to the wall of the room, or placed on a stand made for the purpose, resting on a table or office desk. It also differs from all the others in presenting its diaphragm vertically to the sound waves, instead of horizontally or at a small angle with the horizontal. One of the cases contains the Blake or

Edison microphone described at p. 255, and shown at

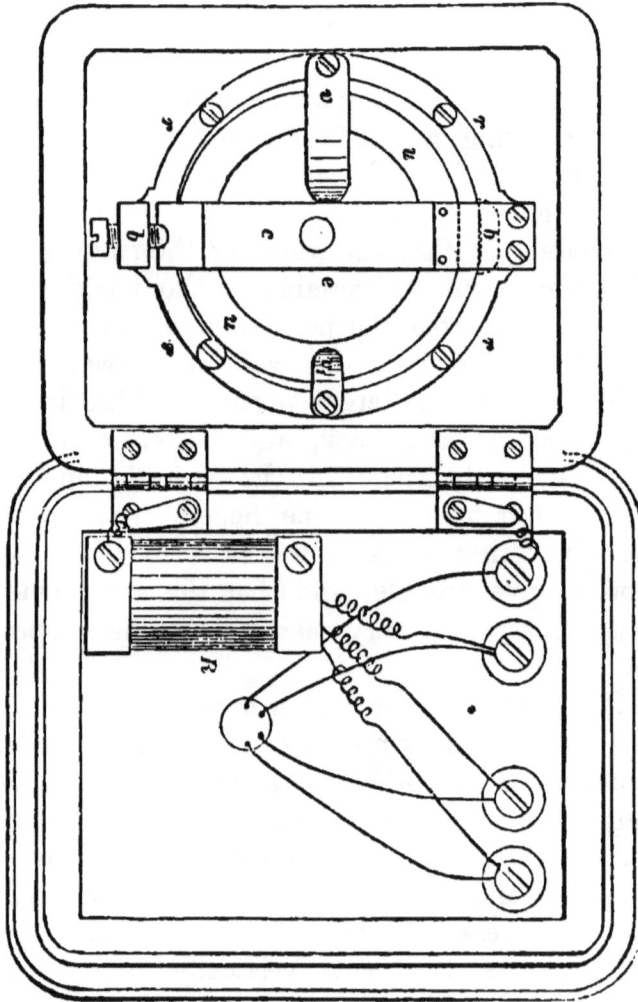

Fig. 115.—Interior of Blake Transmitter, showing Back of Microphone, Induction Coil, and Connections.

Fig. 109, with the induction coil.   The other case contains the switches, the ringing key, and sometimes the call bell.

The switch is perhaps the simplest and most beautiful arrangement that has yet been devised. It consists of a stout brass disc, carrying in its periphery four pins, two of which switch the line from bell to receiver circuit, and the other two close the microphone circuit. The disc turns on an axle in its centre, as a wheel, being impelled by the weight of the telephone receiver on the one hand, when depending from a hook placed eccentrically on the disc, and on the other by the usual opposing spring.

Facing the four pins, are four straight steel springs, in connection with the different wires as described. As the disc turns on its centre, it brings one pin above in contact with its spring on one side and one below on the other side, and when it turns in the opposite direction, one set of pins leave their respective springs and the other set of pins engage with those opposite to them. It will be noticed that this arrangement combines several advantages. The contacts all keep themselves clean, by rubbing together each time they engage, and there need be no undue strain upon the steel springs nor danger of contact when not required, as the pins leave them quite freely, by reason of their forming part of the periphery of the disc, to which the springs are tangents.

The ringing key in this case is also in the form of a small push, the springs and contacts being very much as in the Gower-Bell combination instrument, except that they are inside the case, instead of behind the back-board. The connections of the instrument, however, are principally in the back-board; and in the earlier instruments the hinges of the case formed part of the circuit.

In other forms a small magneto-electric machine is used for calling, the whole of the switches, microphone, etc., being in one case. Attached to the same case is a writing desk, with a small cupboard underneath, in which is placed one No. 2 Le Clanché cell, the apparatus being arranged to work with that.

The Blake combination speaks remarkably well when everything is in order; and it has the advantage that one has not to bend over the instrument either to send or receive; in some instances it being possible to send a message, and receive a reply, seated at a table several feet from the mouthpiece of the instrument.

In the author's opinion, however, it is not equal for all-round kwor, to the Gower-Bell combination, nor to the Crossley or Johnson—mainly, because its microphone is so delicate. It has been already explained, in connection with the merits of the Johnson shunt coil and the Crossley and Gower and Ader microphones, that these latter possessed the advantage of having more than one branch circuit within the microphone itself, so that any excess current is divided between a number of possible sparking places; and that further, if one contact breaks or gets clogged with dust, or burns itself into firm connection, the instrument is not disabled, the other contacts receiving and transmitting the sound-waves.

If this reasoning be correct, the Blake, with only one contact, must be more liable to derangement by sparking, and must require more frequent attention ; and this is found to be so, according to the experience of those who have had opportunities of judging.

In the author's opinion, also, the form of the microphone renders it liable to derangement from changes of temperature, from loud noises, and other causes that will not affect other instruments.

Figs. 116, 117, show Prof. Silvanus Thompson's complete instrument. Fig. 116 being the exterior appearance of one that is designed for long lines, and

Fig. 116.—Prof. Thompson's Valve Microphone Transmitter with two Receivers. The long-distance instrument.

T

Fig. 117 for shorter ones. Fig. 118 shows the exterior of an earlier instrument in which all the switching was performed by one hook switch, of a duck's-bill form.

The apparatus includes the valve microphone already described, the carbons being at the upper end of a wooden tube and covered by a glass; the lower end of the tube forming the usual mouthpiece for the reception of the sound-waves, and being bent at right angles as shown.

The working parts of the instrument are contained in the case in front, the bell being on the base-board, underneath the microphone. In the earlier instruments it was on the top of the switch case.

The ringing key is similar to that of the Blake, and is placed either in front or at the side of the case.

In this instrument, it will be remembered, the receivers have to be excited by a battery current, and the switches therefore have the additional work thrown on them of making and breaking this battery circuit, as well as connecting the bell,

Fig. 117. — Prof. Silvanus Thompson's Combined Apparatus for Short Lines.

the telephones, and the microphone.

Prof. Thompson uses no induction coil with the instrument for short distances; and thereby, as will be seen, greatly simplifies the construction.

For the long-distance instrument, the connections are somewhat complicated. From the line terminal of the instrument, the current passes as usual to the ringing key and the hook switch, and then divides between the battery, the full strength of which is employed, the two receivers in parallel circuit, the secondary coil of the induction coil, and the coils of the bell; all the branches

Fig. 118.—Prof. Silvanus Thompson's Earlier Combination Instrument, consisting of the Valve Microphone Transmitter, with two Receivers, and Duck's-bill Switch.

uniting together at the earth terminal of the instrument.

As in other instruments also, a portion of the battery only is used for the microphone, one part of the microphone circuit being connected to the earth terminal. From the battery, a wire leads to the primary wire of the induction coil, which, however, is in duplicate ; and on passing from the induction coil, the microphone circuit divides into two branches, one branch consisting of the microphone itself, and the other a small shunt coil, which performs the same office as the shunt coil in the Johnson and the multiple carbons in Crossleys, the Gower, Ader, etc.

The above arrangement, as will readily be understood, necessitates considerable complication in the switches, and also a special arrangement for the bell. When the bell is required to ring, for calling attention, the trembler contact is left in circuit; when it is not required to ring, the trembler contact is cut out of circuit; so that when the telephone is in use, the armature of the bell-coils is held up to the magnet. The object of using the bell-coils in this way, is stated by Prof. Thompson to be, to act as an induction plug ; and, as it were, to steady the current which is working the receivers. The author has not found any difference, however, in the received speech, when the induction plug has been dispensed with. The duck's-bill switch that was used in the earlier instruments, has five contacts to make each time the telephone is in use, and one to make when not in use. The lever is pivoted at one end, and held down by a strong spiral spring, so that when the telephone is not on the hook, or rather in the duck's bill, it presses against a bent spring underneath it. Two steel pins pass through the body of the lever, insulated from it, and four brass springs, bent slightly out of the vertical, engage these pins; making contact with them when the telephone receiver is in use, and disengaging from them when the receiver handle is in the duck's bill. The outer end of the lever, which projects from the case, forms with another brass arm, an aperture something the shape of a duck's bill, and when the receiver handle is placed in the mouth of the duck, so to speak, the upper part of the bill is forced up, the lever inside breaking the five contacts already described and making contact with another bent spring above.

The author has no experience up to the present with the short-distance instrument; but, for the reasons already mentioned, he would predict that speech would not be so clear, the articulation not so good, without the induction coil, as with it.

He has heard remarkably good speech from the long-distance instrument; at times quite equal to that of the Gower-Bell combination; but it is not so sure. In order that one may obtain good loud speech and clear articulation, it is apparently necessary to use large battery-power; that is to say, large cells and a comparatively large number of them.

Eight of the largest sized Le Clanché or Sulphur Sal-ammoniac cells are, so far as the author's experience has gone, always necessary for the receivers, even on the shortest lines; and three, four, and even five for the microphones; only two, or at most three, being used with other forms, except where, as with G. P. O. Gower-Bell instruments and United Telephone Co.'s, the induction coils are wound to work with four cells and the winding varied with the distance.

The increased number of switches would naturally also be additional possible sources of failure; and it is found therefore that though this instrument is capable of, and does very good service, it requires more attention than the Gower-Bell, Crossley, or Johnson. One serious drawback to this instrument is, that it is very difficult to arrange more than two stations on one line, if trembler bells are used. If you have three or more, the bells of the stations whose instruments are not in use will be ringing if they are tremblers, and causing a clatter in the receivers of the stations that are trying to speak.

In the whole of the telephone transmitters that have been described, an induction coil is used, for the reasons already given, viz., because it gives clearer articulation.

With the Hunning and the Moseley, however, no induction coil was used; the Rev. Mr. Hunning, the inventor, holding a contrary opinion on this point.

With this instrument, therefore, one or two receivers are used, and only one switch is required besides the ringing key. In fact, even that one may be dispensed with, as the fact of connecting the receiver and transmitter and battery to line, rings the bell at the other end, if the battery is strong enough to ring through the resistance of the local receiver and transmitter.

Where two receivers were used, the transmitter was held in a brass crutch attached to the switch board.

This form of apparatus has the decided advantage that electro-magnetic receivers can be used without, as in other cases, the necessity of complicated connections; but the results are never so good; and even the speech transmitted by these instruments themselves is greatly improved, according to the author's experience, by the addition of an induction coil. Further, though speech is fairly good with electro-magnetic receivers, it is always best with magnetos; that is, with instruments having permanent magnets.

In use, the microphones of the Hunning transmitter frequently get caked or set together, causing blurred · speech; but this is quickly remedied by shaking the instrument sharply.

This defect is not so apparent in Moseley's transmitter; but otherwise the two instruments behave very much alike.

In the Swinton telephone transmitter, the microphone, as already explained, is unprotected, except by the metallic frame in which the carbon pencils are held. The remaining parts of the apparatus are, the backboard upon which the whole is mounted, the bell, the switch that connects and disconnects the telephone from the line, and the ringing key—this last being in the same form as that of the Blake and Thompson instruments.

Switchboard for Two Receivers without Microphone.— Where the former are of the magneto form, all that is required is, that the line shall be switched from the bell to the telephones by the operation of taking the latter off the hook, and that a ringing key shall be available for sending a current through the bell at the other end. Any arrangement will answer that properly fulfils these conditions.

Where electro-magnetic receivers are used, the switch must connect the battery to the line through the receivers when the latter are taken off the hook.

When two receivers are used without a microphone, they are usually connected in parallel circuit; that is to say, the current divides between them. The series arrangement can be adopted if preferred, but the parallel arrangement generally works best. One receiver is used as a transmitter, and the other, or both, as receivers. Three instruments can be used, of course, by providing a rest for the third, to avoid the necessity of removing one from the ear.

The arrangement of two or three receivers is only efficient for very short lines, such as for communicating between different parts of dwelling-houses, etc., and not then if either station be in a noisy place.

Telephone Circuits.—To connect two or more stations together, we require a complete set of instruments at each station—a transmitter of some one of the forms described, one or two receivers (two are always better than one), a call-bell, a battery or small magneto-electric machine to operate the call-bell at the other

end, a battery for the microphone (forming usually
part of the calling battery), switches, and a ringing
key. In addition to these, an insulated wire is required for connecting station to station, and either
earth or a return wire to complete the circuit.

Where the wire can be fixed overhead, a naked galvanized iron wire is used, usually No. 11, though the
telephone companies use a stranded wire consisting of

Fig. 119.—Johnson's Fluid Insulator.   Fig. 120.—Fuller's Insulator.

three No. 16 wires. The line wires are supported
either by the " Z " insulators shown in Fig. 95,
p. 237, or some other form designed to give a higher
degree of insulation, such as Fuller's, shown in Fig.
120, or Johnson's fluid insulator, shown in Fig. 119.

For terminating line wires, the shackle insulators
shown in Fig. 86 are used. Telephone wires should
always be terminated, or shackled off, as it is technically expressed, on either side of streets, roads, etc.

Over towns, great care is required in fixing telephone
wires, as serious accidents may result should a wire
fall in the street. In order to minimize the mischief
that a falling wire can do, streets should always be
crossed as nearly at right angles as way-leaves will
allow ; the wires should be kept off the main thorough-
fares as much as possible, and should be carried as
high as adjacent buildings will
permit. It is usual to attach
the insulators of telephone
wires to the chimneys of
buildings on the route, where
permission can be obtained ;
using for this purpose either

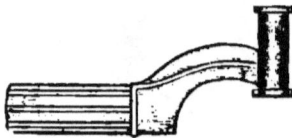

Fig. 121.—Wall Bracket.

a corner bracket, consisting of a piece of iron bent to
the shape of a chimney and nailed to it; the fluted
bridge or wall brackets shown in Fig. 121 ; or, better
than either, especially where the chimney is not of
the strongest, a stout wire band passing twice round
the chimney, the two portions of the band lying on
different courses of bricks.

The corner brackets are nailed to the chimney by
the large clout nails used for pole brackets, the nails
being driven into the plaster.

The bridge or wall brackets can only be used when
the brickwork is very substantial, and when the strain

is either towards the chimney or at right angles to the line of the bracket, and never where there is much strain, such as in the case of a long span.

Corner brackets also, though very convenient and inexpensive, should only be used when the resultant pull is towards the chimney, or at least at right angles to one arm, and then only if the span is short.

One very great trouble that arises with telephone wires in towns, is the humming that frequently accompanies them. It can be heard at any telegraph pole on a country road, and is often very loud. Sometimes it is due to the wind passing across the wire, setting it in vibration; sometimes, apparently, to vibrations caused by heat. In either case, the result is apt to be very troublesome to the inmates of the houses to which it is attached.

The trouble is, however, not difficult to overcome. Make your spans as short as possible. The humming and howling are always worst on houses or poles forming rests for wires that stretch some distance before the next rest is reached, and particularly when the space passed over is open to the prevailing winds. Next, you have to break the acoustic circuit between the wire and the chimney. Terminate your wire on opposite sides of the chimney, using separate wire bands, and keeping these as far apart on the chimney as possible, and make the connection between the two parts of the wire by means of a piece of wire well covered with india-rubber.

Often this is sufficient; but if not, replace the ordinary shackle insulator by one of india-rubber; the acoustic circuit will be completely broken, and the humming will give no further trouble so long as the india-rubber remains sound.

When the rubber perishes, as even the best does with exposure to the atmosphere, it must be replaced, or the acoustic trouble will begin again.

Another cause of trouble with telephone wires over towns is, the smoke and acids that rise over the town rapidly rust the wire; the difference between the life of a country wire and of a town wire being at least as two to one. Good wire, well galvanized, arrests the action for a time, but only for a time. The zinc covering is soon used up, leaving the iron exposed beneath. Often, too, a flaw in the galvanizing will allow chemical action to commence and to go on underneath the zinc.

Wires over towns, especially smoky ones, should be overhauled every autumn, and any portion that is getting thin renewed, as the winter storms will be sure to find out weak places, break the wire, stop communication, and possibly do some damage to the roofs as well.

This difficulty may be got over in several ways. Larger wire may be used, giving, of course, more substance, but at the same time larger surface for chemical action, and for the clinging action of the snow that is so troublesome at times.

A better plan is to use No. 11 wire, covered with a complete envelope of braided yarn, saturated in composition that will resist acids.

If the braiding is well done, it very materially lengthens the life of the wire, though it also materially adds to its cost.

Another plan is, to use copper or phosphor bronze wire, it being well known that most chemical elements that are found in the atmosphere of smoky towns have a far greater affinity for both iron and zinc than for copper.

Where the line is long, this has the advantage also of reducing its resistance, which is of importance; though again it adds to the cost if either the protection from chemical action or the reduction of line resistance are to be of any value.

One thing should always be avoided. Never fix wires with welds in them over towns; use wire drawn in one length, or cut the weld out and make a joint. Never allow a kink in the wire to remain. If serious, cut it out. The wire will always part at welds or kinks.

Snow is occasionally a great destroyer of telephone lines; but it would appear to be only once in some eight or ten years that we get a storm in this country that combines in itself the necessary qualifications for serious damage. When, however, it does possess those qualifications, the destruction is very serious indeed.

As a rule, the snow either falls harmlessly off the wires, or if it does lie sufficiently thick, and for a sufficient time, its effects are felt most on the insulation of the line, the weight of the feathery particles not being usually very great. But in certain storms, fortunately rare, the snow appears to settle on the wire and freeze there, forming a bed for fresh snow, which also freezes, so that the size of the wire goes on increasing ; and, moreover, it is a compact mass of ice, presenting a large surface to the wind, the result being that miles upon miles of strong, well-found telegraph and telephone lines have been wrecked in a few hours; there being, of course, in addition to the weight of the wire and ice, and the wind pressure on the enlarged surface of the wire, the enormous contractile force due to the very low temperature—this contractile force acting not only on the wire, but on its icy skin. All these forces act in the same direction ; and the result has been that the heavy storm which was felt all over the kingdom a few winters back is stated to have cost the United Telephone Company £30,000 in damage done to wires, fixings, buildings, and incidentals, over London alone ; while the Great Western Railway, the General Post Office Telegraphs, and others, all suffered very severely.

For connecting the outer line wires to the instruments, or for running underground, where that can be done, No. 16 or No. 18 copper wire should be used, covered to No. 7 or No. 6 with either gutta-percha, india-rubber, or Callender's compound. The larger wires are preferable, and there is not much choice between the three forms of insulation, as all perish with exposure to changes of weather.

Care should be taken, in fixing these wires, that the insulation covering is not damaged, especially where any wet drips. Do not lead the wire over the sharp edge of an iron water-shoot. Protect the wire by boarding if possible.

For the inside of offices, etc., and for connecting to the battery, No. 20 and No. 22 copper wire will do, covered with india-rubber, gutta-percha, or Callender's compound, and then laid over with cotton, the same wires that are used for domestic electric bells. The larger the wire and the thicker the insulation, the longer it lasts without repair ; but a No. 20 for battery wires, and No. 22 for other connections, usually answers very well for private telephone work. It is generally necessary to staple the wires used for connections to telephones, in offices, etc. This should be avoided if possible ; but where no other plan of securing them in their places can be arranged, be careful to staple each *battery*-wire by itself, and to see that the staple does not touch, either another wire or another staple holding another wire.

It is wisest to staple all the wires separately; but as it is difficult to make them look neat, if this plan is adopted, those used for general connections, as to bell, line, earth, etc., may be stapled together without much danger, if care is taken. Ot course, if the line and earth wires, say, are stapled together, and the staple cuts into both, making connection between the wires, the instruments are cut out by the short circuit formed by the staple. But with well-covered wire, and careful stapling, there is not much danger of this ; and if it should occur, the most that could happen would be a stoppage that any intelligent line-man would discover and put right in a few minutes.

If, however, one of the battery wires should be stapled with another, and a connection be made between them, it may bring a serious strain on the battery; and further, if moisture be present, will assuredly part them.

Induction and Leakage.—Two other sources of trouble with telephone lines are induction, and leakage from wire to wire; causing a clattering in the telephone receiver when powerful currents for telegraph instruments are passing in adjacent wires, and reproducing telephone messages passing in them.

The cause of this is twofold. Where wires run for miles together on the same poles, a leakage path is formed from wire to wire at each pair of insulators, by way of the moisture that usually condenses on the cold surface of the insulator, the iron bolt of the insulator, the iron bracket, and the pole itself, if damp. It will be evident, in this case, as in that of the engine-plane mining signal, that though the resistance of one such path may be very great, the resistance of a large number will be comparatively small, so that a current passing in one wire can find its way into any others on the same line of poles, and may operate the instruments attached to them. If any two wires run together for a sufficient distance to reduce the resistance of the leakage path between them, the combined resistance of this, the path, the wire, instruments, and earth on either line may be small enough to allow a

current to pass through them from the other line of
sufficient strength to work the apparatus; and as telephone receivers work with very low currents indeed,
it will easily be understood that on long lines there is
considerable trouble from this cause.

There is also another form of leakage from wire to
wire, viz., by way of earth. Where, as frequently happens, a number of telephone lines run into a town, and
are connected to the gas or water service, especially
if the joints of the branch pipes to the mains are not
very good; or, in other words, the resistance of earth is
comparatively high; a portion of the incoming current
may find its way out again through other wires connected to the same earth, and complete their circuit
through another path.

Nor is it necessary, for earth leakage, that the second
wire into which a portion of the original current passes,
should be connected to the same metals in the earth.
Provided that the path for the current from the receiving station to the sending station (or one of them,
as there may be several), crosses the path of another
wire terminal, it will naturally include that wire in the
return circuit. It must not be forgotten that earth
circuits obey Ohm's law just as every other circuit
does; and further, just as any conductor looped to another conductor carries a portion of the current, in accordance with the law of branch or divided circuits,
so does each portion of the earth's crust take its share
in the work of transmitting the return current, often
taking it miles round to bring it home; and so each

U

conductor connected to it by an earth plate, and perhaps connected to earth by another plate at the other end of the line, carries off a small portion of the return current, and may thus divide it with another wire on the same poles and going to another earth, and yet again with other wires that have terminals in the path of its own earth. Thus it may easily come to pass that a portion of a telegraph current sent over a few miles of wire, say in Cornwall, may, on its return journey, pass up to Glasgow and down again.

The above is merely intended to show how it can be possible for a current passing in one wire to appear in another with which it has apparently no connection. With comparatively coarse instruments, using powerful currents, such as the single-needle and the old A.B.C. telegraphs, leakage currents could only affect their own line, by reducing the strength of the current at the receiving station; but with such sensitive apparatus as telephone receivers, which necessarily respond to very small forces, the trouble is more to get rid of these stray currents and stray messages; or rather, in the latter case, to prevent the message on one wire finding its way into others.

It will be easily understood, of course, that each time the current splits off in the manner described, it becomes weaker, and its action upon a telephone receiver less; so that for practical purposes one has not to go beyond the immediate earth connections and line leakage connections of any wire.

At the same time, it may be mentioned, that in

America, a message has been heard on one wire six miles from the wire in which it was transmitted, there being no connection between the two except by way of earth.

The other cause of the clattering in telephone receivers connected to wires running in the neighbourhood of other wires in which telegraph messages are passing, and of telephone messages being reproduced on neighbouring wires, is the peculiar phenomenon known as Induction.

The phenomena are fully described on page 40.

For present purposes it will be sufficient to remember that where two wires run side by side, as telephone wires do on poles; when a current passes in one wire, at the instant of its passage, a current is generated in the other wire, in the opposite direction, by induction; and when the first current ceases, a current rises in the second wire in the same direction as that which was passing in the first. As already explained in connection with the Induction Coils that are used in telephone transmitters, neither an actual generation nor an absolute cessation of current are necessary for induction to take place, a variation of current is sufficient; but of course the greater the variation, the greater the induction, and thus the sharp reversals of the comparatively high-tension currents used with single-needle, Morse, and other telegraph instruments, give rise to sharp clicks in the telephone receiver, which drown any spoken message, unless special arrangements are made to overcome it.

Telephone messages, however, are faithfully reproduced upon neighbouring wires, and often with as much distinctness as if the two wires were one; though it is not clear whether this result is due to Induction altogether, to leakage, or to both; probably to both.

Then, as to the remedy for these complaints. First, let it be noted that there are two complaints to be dealt with. One is, that,—owing to the clattering caused in telephone receivers attached to wires running on poles with other wires, by Induction and leakage from the wires carrying telegraph currents,—it is, or was, often impossible to hear the telephone message. It was as though some one was hammering on a tin can close to one's ear while a second person was speaking. In fact, it was easier to hear a telephone message in a large ironworks, where huge steam hammers were working, than to hear on a line running for any distance in the neighbourhood of telegraph wires.

The remedy for this complaint lies in increasing the power of the telephone current entering the receiver; and this may be done by increasing the E.M.F. generated by the induction coil, either by winding the coil itself specially to give a higher E.M.F. with a given battery; or, if it is inconvenient to alter the instrument, by increasing the battery power in the microphone circuit.

With telephone apparatus using Induction coils, the author has never failed to secure good speech over lines up to thirty miles, running on poles with upwards of thirty other wires using telegraphic instruments, provided the Induction coil was properly constructed. Two receivers, or one with two tubes, such as the Gower-Bell, were always used, and a sufficient number of cells employed in the microphone circuit. The effect of adding cells to the microphone circuit of a Gower-Bell combination instrument, where the induced clatter has drowned speech, is very striking.

With direct-current instruments, the equivalent remedy is to employ more battery cells in the line wire; but, so far as the author is cognisant, no striking success has been obtained over long distances with these instruments, when opposed to Induction and leakage from other lines.

Another very beautiful remedy, which has been introduced on some of the Continental telegraph lines in order to use them for the transmission of both telephone and telegraph messages on the same wire at the same time, is that designed by M. Van Rysselberghe. He places at each end of the line, between the line wire and the telephone receiver, an electrical condenser; that is, an apparatus for storing electricity electro-statically. It will be noticed that the telephone circuit is actually broken at each end, by the condenser, so far as the passage of a current is concerned; but that the variations of E.M.F. arriving at one plate of the condenser are faithfully reproduced by electro-static induction at the other plate, and thence in the telephone. The telegraph instrument being on a branch circuit connected to line beyond the condenser, receives its own current, but cannot work through the condenser; while, on the other hand, the resistance of the condenser is so high that the telegraph current, for practical purposes, cannot pass through it.

The writer is not aware if this plan has been tried for stopping induction between wire and wire; but it is certainly probable that the extra work given the induced current, or the leakage current, by the condenser, should weaken its effect upon the receiver.

The other trouble produced by induction and leakage is the overhearing of messages. It is obvious that this is a very serious matter. Where two wires running together are owned, one say by a firm of cloth manufacturers, and the other by an iron firm, the inconvenience is not serious. In very damp weather, with the insulation of the wires very low, it may necessitate one firm's waiting till the other has done speaking; but no serious harm is likely to follow if the iron people are advised of the price the cloth people are going to ask for green meltons; nor if the latter discover the latest price of best pig iron. But, where there are several firms, all in one line of business, whose wires run on the same poles, as in the case of a number of collieries who ship their coals at the same port, the inconvenience may be very serious indeed. The remedy for this,—and there is only one, so far as the author is cognisant,—is to use a complete metallic circuit, thus displacing earth and one source of leakage. Further, the return wire is run as close as pos-

sible to the line wire, and with a twisting motion, each
wire circling round the other, as shown in Fig. 122.

The object is to neutralize the induction. It will be
obvious that if two wires, forming the line and return
are exposed to induction from the same primary wire,
the net result on the circuit so exposed will be *nil :*
the *two* induced currents opposing and neutralizing

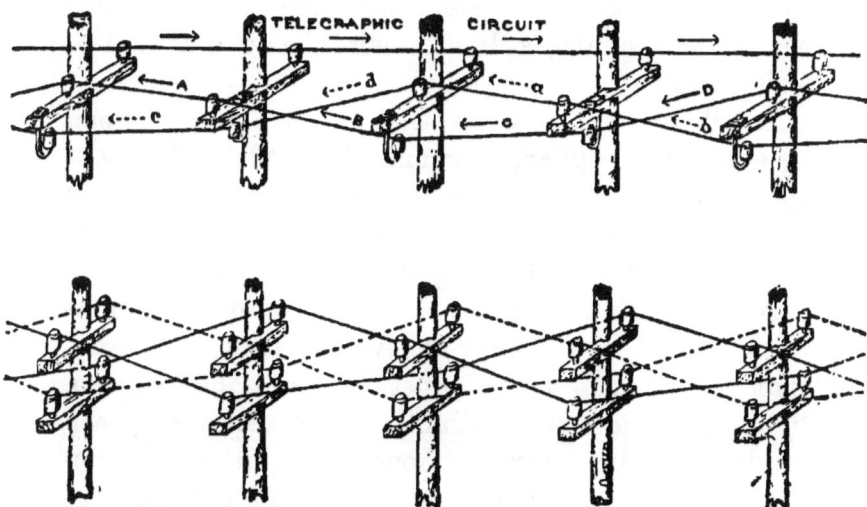

Fig. 122.—Arrangement of Telephone wires to neutralize Induction
from wire to wire.

each other. But even this has not been found to be
completely effectual; as, owing to differences of position, there may be a difference in the E.M.F. generated
in one half of the wire from that generated in the
other half; and this difference, if sufficiently great,
will operate the telephone receiver.

The plan shown above has, therefore, been adopted,

in order that the relative position of the two wires may be the same with reference to other wires throughout the line, and that the insulation of the line wire from the return may be efficiently carried out, which would not be always an easy matter where they are very close together. But here a mechanical difficulty has crept in. Telephone lines fixed on this plan take up more room on the poles, and are more difficult to maintain than when single; a broken wire being a source of serious trouble.

It will be seen, however, that the plan, where it can be efficiently applied, overcomes both the difficulty of speech being drowned by the clatter of the induced currents from telegraph lines; and also that of users of neighbouring wires hearing each others' messages.

Another plan was tried with Hunning's transmitter and Cox-Walker's receivers. It was thought, that if the effect of induction could be reduced, by diminishing the sensibility of the receiver, speech might still be heard without the clatter of telegraph currents, and without danger of overhearing. For this purpose the ordinary Hunning-Cox-Walker combination was used, and the electro-magnets of the receivers were pulled back by the regulating screw until induced currents no longer operated them. Then, when this point was attained, the battery power was increased until the speech, which had also disappeared or become very faint, reappeared in its full vigour. Unfortunately, when speech reappeared, induction did also; the apparatus that could be operated by the currents representing the one, responding readily to the other. Another important point that came out in the course of the experiments was, the immense superiority of the Gower-Bell combination, with its induction coil, over the direct current arrangement. Two Gower-Bell instruments were connected to the same wire, and under the same conditions as those ruling with the Hunning-Cox-Walker apparatus, the line being seventeen miles long, and running on the Taff Vale Railway poles with some fifty others, most of the latter using single-needle telegraph instruments.

Long after speech had ceased to be intelligible in the Cox-Walker electro-magnetic receiver, when using Hunning's transmitter, it was distinct if the Gower-Bell transmitter was used, though the resistance of the Hunning transmitter was interposed. In fact, it was difficult to regulate even the Cox-Walker receivers, so that they failed to receive when the Gower-Bell transmitter was used for sending.

Multiple Stations.—Where more than two stations are to be connected, the simplest plan is as follows. At each station fix a complete set of apparatus, including a transmitter, one or two receivers, a call-bell, and a battery or magneto call and connecting, and earth wires. Run the line wire to the neighbourhood of each station, merely leading the covered wire from it to the line terminal of each transmitter; so that the apparatus will form branch circuits, the line wire being common to all, and each branch leading to earth.

Then, when one station calls, the bells at all the others ring; and by giving each station a number, any particular station can be called at will.

The author has arranged as many as six stations together successfully on this plan. One drawback to the plan, however, is the fact that all the stations can hear what any two may be saying to each other. On the other hand, this has its advantages, as a head manager can summon all his subordinates to a conference by telephone at any time, though they may be miles apart, and the conference can be carried on practically with as much ease as if they were seated together in one room.

It must be noted, however, that it is not easy to arrange this plan of multiple stations with direct-current apparatus, such as Hunning's, Moseley's, or Swinton's, or with those in which a powerful battery is connected to line when speaking, as in Silvanus Thompson's; and for the reason already mentioned, viz., that the bells of the other stations may be ringing the whole time that any two are attempting to speak, the result being a continual clatter in the receivers, which effectually drowns speech. The clatter is due to the variation of the exciting current of the receivers by the breaking of the other bell-circuits, and the induced currents generated thereby. Connecting the batteries so as to oppose each other will not help the case much, as, though batteries *may* be of exactly equal E.M.F. and resistance when first connected up, they do not remain so long; and even if they did, when the batteries at stations A and B are connected to line, the bell at station C, with its earth connection, forms a circuit for both.

One method of overcoming this difficulty is, by using single-stroke bells, and allowing the hammers of the bells of other stations to remain up while two are speaking. If this plan is adopted, the resistance of the bells should not be low, as otherwise they may seriously affect the received speech by shunting a portion of the working current.

Single-stroke bells, however, are not so well adapted for calling attention as the continuous sound of a trembler bell; and therefore, if some other plan can be found, it is better.

In one installation that was fixed by the author's firm, they were successful in working four stations with direct-current instruments, and without any trouble with clattering, such as described above. The secret of success seemed to be, that the bells were constructed to work well when connected directly to the battery at the calling station, through the ringing key of the latter; but would not work with the small current which fell to their share when divided with two other bells and the receiving apparatus at the third station; the transmitting apparatus at the fourth station being common to all the branches. Even in this case, if the batteries were very strong, one or other of the bells did give an occasional clatter, but it never lasted.

Faults in Telephones.—Telephonic communication between two places may fail, either from one station being unable to call the other, or from failure to transmit or to receive speech.

If either one of these causes be present, the whole apparatus is useless for practical purposes, since, if a station is either unable to call or unable to receive a call, it cannot communicate except at uncertain times; and if either station cannot send or cannot receive a message, it is not of much use the other being able to, as it takes two to make up a conversation.

Therefore, whoever is in charge of telephonic apparatus, or whoever may be called upon to put telephonic apparatus in order, would do well to follow the rule already given, and spend a few minutes in making careful inquiries as to what portion of the apparatus has failed. A user will generally only know, at first, that he cannot communicate; but a few careful questions will usually elucidate such facts as—that they can speak all right if they can only call attention, which shows at once that it is the calling part of the apparatus which is out of order; or, it may be they can call all right, and one end can hear the other, but the other end cannot understand; and so on.

If of two stations, A and B, A cannot call B, while both can hear, clearly the fault must be in A's battery, ringing-key, or connections, or in B's bell or connections, and *vice versâ*. If A can hear B, but B cannot hear A, the fault must be in A's sending apparatus, or in B's receiving apparatus, and *vice versâ*.

If neither can speak or hear, but each can call the other; the cause may be in either sending apparatus, or, more probably, it is a break in the receiver circuit. If neither station can call, nor hear, it is either a break in the line, faulty earth, or both batteries are exhausted.

In any case, always commence by testing the battery at the station you first arrive at. The battery is the part of the apparatus most likely to fail in every case, therefore always make sure of that before proceeding further. Then, if it is a case of B's bell not ringing, and you are at station A; carefully examine the wires and connections between the battery, the key, and the line. A break in the wire leading from the battery to the transmitter battery terminal, or in the wires connecting that with the battery stop of the key, will prevent a current being sent to line. Such a break might occur at a staple, at a terminal of the transmitter, either outside or inside,—wires often breaking off short at terminals,—at a soldered connection to the battery stop of the key, or any other soldered connection, or at a kink in the wire if it has been bent sharply. If no break can be seen, or detected when the wire is pulled, test from the battery stop of the key to earth with the short circuit of the galvanometer. The deflection should be the same as at the battery.

If still no fault appears, disconnect the line wire from its terminal and insert the galvanometer, and then press the ringing key as for calling. Do this with the long and the short circuit of the galvanometer interposed alternately, and carefully note the deflections. If there is no deflection with either galvanometer circuit, disconnect the line wire and the battery wire, and connect them together through the galvanometer. If still no deflection is obtained, the fault is probably at the other end. If a deflection is obtained with the battery and line wires connected directly together, and none when the key and hook switch are in circuit, the fault is in the instrument. Connect the line wire to its terminal, leave the battery wire connected to the galvanometer, and with a wire from the other terminal of the galvanometer connect successively to each point in the path of the line calling current, from the line terminal to the ringing key. When two points are found, at one of which, when connected to the galvanometer wire, a deflection is obtained, and at the other none, or the deflection is greatly reduced, the break will lie between them. If the fault appears to be at the other end, either leave some one at one end to send a call signal when you are at the other station, or connect the calling battery wire and the line wire directly together. The former method is preferable, if you have some one to leave who can follow your instructions and will do exactly as you wish ; as, if you connect the battery wire and the line wire together, you will have to come back to the first station to disconnect these and to connect them to their proper terminals. On the other hand, clerks and others who use the instrument have not always the time to give to your requirements, and do not always clearly understand what you require.

In any case, make some arrangement to have a

current on the line, in a similar manner to that described for shaft signals, and proceed to find this current at the other station.

First disconnect the line wire from its terminal, and connect it to the earth wire, through the galvanometer, observing the deflection both on the long and short circuits; the deflections of these two forming a very good guide to the nature of the fault, when a little experience has been gained. If no current is obtained, work back to the iron overhead wire; but this will only be necessary in the case where no current passes either way. Where the bell rings one way, it is usually proof that the line is perfect up to the line terminal of each instrument. Assuming therefore that a deflection is obtained with a current on the line from the other station, when line is connected to earth through the galvanometer, but that the bell will not ring; examine the latter carefully; see that there are no wires broken either inside or outside the bell-case, that the insulated contact screw against which the spring of the armature presses, is making firm contact with its spring, and also that the spring leaves the contact when the hammer moves to the bell. See that the earth connection of the bell is perfect. Examine the hook switch inside the transmitter, the connection to the spring of the ringing key, and the back contacts of the hook switch and ringing key.

If nothing is discovered, such as a broken wire, dirty connection, or dirty contact, proceed with the test. Connect the line wire to its terminal, leaving the galvanometer connected to earth on one side, and the other wire free. With this free wire touch each point, in succession, in the path of the current from the line terminal to the earth side of the bell, as the inside of the line terminal, the fulcrum of the hook switch, the back stop of the hook switch, the lever of the key and its back stop, the terminals of the bell, etc. As before, a deflection will be obtained through the galvanometer at each point, until the break is passed; and the latter will lie, as already so often detailed, between the last two points where tests were made.

If neither bell rings, proceed exactly as detailed above; but in addition, as the fault may be in the line wire itself, test the latter in the same way, putting a current on and connecting to earth through a galvanometer, until the two points are reached, at one of which a deflection is obtained and none at the other.

It is not always an easy matter to test from the line wire to earth, as the latter may be difficult to obtain. On the other hand, it is not often that the fault is in the line wire; and if it is, it will most assuredly either be a defective joint between two lengths of iron wire, between the iron wire and the copper leading-in wire, or a break in the latter.

A careful examination of all joints should discover the fault, if careful tests at each end show the fault to be in the line wire; and it is then a simple matter to repair it. A detector galvanometer with which the tester has had some experience, is of immense value in tests of this kind; as, by taking the two tests he can judge of the nature of the fault with great certainty, more particularly if he has made previous tests on the same wire with the same instrument. Thus, assuming a partial disconnection in some portion of the calling circuit. On inserting the galvanometer, the short circuit would probably show no deflection at all; so that if that test alone be depended on, a total break would appear to be present. The long circuit, on the other hand, would show the deflection more or less reduced, according to the nature of the fault; and from the two readings, combined with previous experience, it would be known that probably a dirty contact or a joint not well soldered was responsible for the failure. If neither circuit showed a deflection, the line-man would know at once that he had a total break in the circuit, such as a broken wire, or a contact-piece not touching its fellow.

Again, there are cases where a bell refuses to ring because the current which ought to pass through its coils, passes either wholly or partially by another path. Some one may have accidentally disconnected the bell wire from the transmitter and connected it to the earth terminal, so that the calling current is passing directly to earth instead of through the bell coils. Or, the naked line wire may be touching something which is carrying off a portion of the current before it reaches the bell, not allowing sufficient to reach the latter to work its armature and hammer.

In such a case, the long circuit would not afford much guidance by itself, as it would merely indicate

x

a very slightly increased deflection, and possibly not that, if the battery was not at full strength. The short circuit, however, would tell the tale instantly, by a considerable increase in the deflection. A fault of this kind is exceedingly troublesome to find, unless two are engaged in the work, one remaining at one station with the galvanometer in circuit, noting the deflection from time to time when the key is pressed, and the other going to successive points on the line and disconnecting, as described for Mine Signals. It would be an almost endless task for one man to find a fault of this kind, unless it happpened to be very near one station, or he possessed a very quick eye, and was familiar with every portion of the wire, as he would have to come back each time to note his deflection. With two, however, the fault should only take the time necessary to get to it and to make the successive disconnections and reconnections on the way. Even with two men, however, on a line wire, considerable time may be taken up, if joints have to be disconnected and re-made. Probably one man would go through the line, disconnect at one end, and then carefully pass each part of the wire in review, the other man noting the deflection from time to time until it disappeared. The disconnection of successive sections would be resorted to only as a last measure. Where a Wheatstone bridge is in daily use, a simple resistance test will show the position of the fault approximately; but probably those for whom these pages are written would hardly care to be troubled with such a delicate and expensive piece of apparatus for the sake of discovering a fault that does not often occur. Imperfect insulation of the line wire, say from broken insulators not having been replaced; or from those in use having become filled with dirt, will have the same effect, though usually in a minor degree, as direct contact between the line wire and earth.

The construction and use of the Wheatstone bridge will be fully detailed in the other part of the work.

Faults in Hearing or Receiving.—If Station A reports that Station B cannot hear when Station A speaks, but that Station A can hear when B speaks. Examine the microphone circuit at A. The microphone battery, having the hardest work, most frequently breaks down. If the microphone cells test weak, change them for some of the others in the calling battery; or where a magneto-call is used, renew the microphone battery. Carefully examine all contacts, wire connections, etc., and observe the sound in the receiver at Station A when the microphone board is tapped. If this is clear and loud, the fault is in the receiver at the other end; if not, it is probably in the microphone at station A, which must be carefully overhauled, with its contacts and connections; and the same plan of bridging over successive portions of the microphone circuit may be adopted, that has already been described for other faults.

It has been already mentioned that microphone contacts spark when spoken to, and create dust, which in some cases dulls speech, and in others, gives rise to a humming sound which effectually drowns it. The dust may usually be got rid of by a smart rap on the case of the instrument. There must not be too much rapping, of course, or the instrument may be damaged.

If the fault is in the receiver at the other end, it may be in the diaphragm. With continual work, the centre of the diaphragm, especially in membrane receivers, is sometimes pulled so close to the magnet pole, that its free vibration is impeded, and speech is more or less muffled. With iron diaphragms, turning the diaphragm over is usually sufficient to remedy this. With membrane diaphragms, the magnet must be drawn back, or a new diaphragm fitted to the instrument. A peculiarly troublesome fault of this kind occurred to some membrane instruments that passed through the writer's hands. In making the instruments, some filings appear to have been left inside, quite out of sight. In use, however, these worked out and formed a magnetic connection between the magnet pole and the diaphragm, preventing the latter from vibrating, and completely stopping the speech. It was a very simple matter to remedy this, but it was often very troublesome, as it necessitated a railway journey to accomplish.

It will also sometimes happen with new telephone receivers, that the diaphragms are placed rather too far from the pole; and the inductive effect of the latter is so much weakened that speech is faint. With iron diaphragms this can be remedied by bending the centre of the diaphragm in a little. With membrane diaphragms, the magnet must be brought closer, or a new diaphragm fitted.

Where both bells ring, but neither station can hear, the fault is generally in some portion of the receiver circuit. It may be a wire connection broken from a terminal inside the transmitter, or at a kink within the induction coil, or where it has been soldered. A switch may not be making good contact; or lastly, and most frequently, a wire may be broken inside the flexible telephone cord. This last is a very troublesome fault, as it is apt to be intermittent; that is, it will be on and off within a few minutes. In the early days of telephone work, faults of this kind gave more trouble than almost any other. The reason they are sometimes intermittent is—the flexible wire is composed of a number of fine wires, or in some cases of a fine wire wound transversely on a core of cotton or silk. In either case, the wires in course of time break; but the broken ends being close together, from the construction of the rope, sometimes they are in contact and sometimes apart; and so it will happen that even moving the arm holding the telephone receiver, will make or break the contact, so that speech may be cut off in the middle of a word.

The contact springs of some instruments also frequently give trouble. Where a spring has to work to and fro, and the passage of a current through it to another spring or lever depends upon its always being firmly in contact with the latter, the spring should be made either of very hard brass, well hammered, or of steel; and it should be difficult to make the spring take any permanent position except the one intended for it. Unfortunately, however, it is sometimes found convenient to stamp these springs out of soft brass, which works more easily. The result is that, though at first these springs work fairly well; after a time, depending upon the amount of use they get, they stand back out of their proper position, the spring being gradually taken out of them, and they fail to make contact.

The finding of one of these faults, where neither can hear, though both can ring, is a very simple affair, and consists simply in testing at each station with the battery and galvanometer or telephone receiver. Disconnect the line wire from its terminal on the instrument. Take a wire from the battery wire through a detector galvanometer, and touch successively on the line terminal, and at each connection and each contact, noting the deflection on the galvanometer; or taking the telephone receiver off its hook and noting the click.

It will be obvious that there will be no deflection on the galvanometer, and no click in the telephone, as long as there is a break in the circuit formed by earth, the battery, the testing wire and galvanometer, and the portion of the instrument included between the testing wire and earth; but immediately this break is passed, the galvanometer needle will be deflected.

To find a fault of this kind, test the battery at

Station A and see that it is up to working strength; test through the transmitter and receiver as described; then, if the fault is not found, remake all connections at Station A, as for regular work; proceed to Station B, and repeat the operation.

Telephone Exchanges.—Where the number of stations to be connected exceeds four, and there is likely to be much inter-communication, the simple branch circuit system above described would probably not be suitable; and what is known as a Telephone Exchange becomes necessary.

A telephone exchange may consist of any number of stations, from three upwards, and may be arranged in different ways. Thus, all the lines from each station may be led to one central office; or there may be district offices in different localities, into which the lines from the stations in that locality are led; the station itself being in connection either with the general centre or with another local centre between itself and the general centre. In any case, the arrangements are much the same. There is a complete set of apparatus at each station. At the centre there is also a complete set of the same apparatus; but in addition, there is an indicator, one of whose numbers is connected to each line wire. When any station calls, its indicator shows, usually by dropping a disc on which its number is given. The attendant at the centre, whether local or general, immediately switches his instrument on to the line whose indicator has dropped. If that station merely wishes to communicate with the centre, he receives the messages, disconnects the line from his instrument, and waits for another call. If the calling station wishes to speak to another station, the attendant calls that station and connects the two lines together by means of the switch provided for the purpose, so that the two stations can communicate.

If the calling station wishes to communicate with a

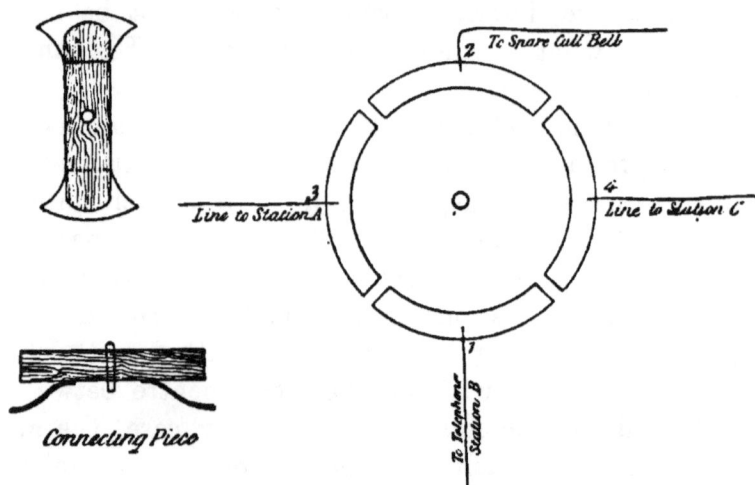

Fig. 123.—Circular switch for connecting three Telephone Stations. When the connecting piece is turned in one direction, say top end to the right, stations A and B can speak, station C is connected to the spare call-bell. When turned to the left, stations B and C can speak, station A is connected to spare call-bell.

station not in direct connection with the local centre, the attendant will call the general centre and connect the caller with that station, and the operation will proceed as before.

For only three stations, where one is only required

occasionally, and it is desired not to hear what the others are saying, a circular switch arranged to connect the telephone at the central to one of the lines, while the other line remains connected to a bell, is very convenient. It will be noticed that either station can call the centre, even when the other is speaking.

The principle of the indicators used for telephone

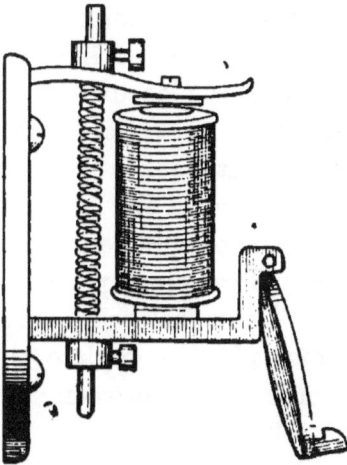

Fig. 124.—Telephone Indicator.　　Fig. 125.—Telephone Indicator.

exchanges, is the same as that of domestic house-bell indicators; but as they have to work over very much longer lines, and to stand uncertainties in the matter of earth connections, leakage, etc., they are made more sensitive, that is to say, more readily responsive to weak currents; and they are also made, or should be, even less liable to derangement from changes of temperature.

Figs. 124 and 125 show the two principal forms of indicator that are used. In each, the action of the current upon the electro-magnet attracts the armature and thereby releases a shutter, which, on falling, reveals the number of the caller.

Switches for telephone exchanges are somewhat varied. For small exchanges a plug switch may be used. Placing the plug in one hole between two pieces of brass makes connection between line and indicator; placed between another pair of brass plates, it makes connection between the line wire and the telephone apparatus. Two plugs placed in two holes between three brass plates complete the connection between the two lines.

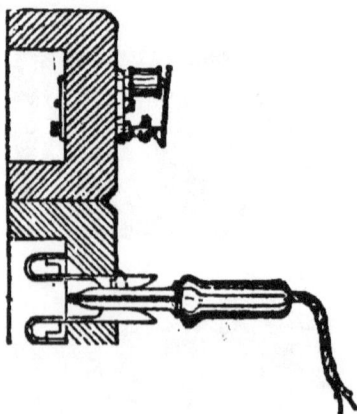

Fig. 126.—Telephone Indicator, with Connecting Jack.

A more favourite plan with the telephone companies is that shown in Fig. 126, where, when at rest, waiting for a call, each line is connected to its own indicator and thence to earth, through the springs shown.

To connect with the telephone, and with each other, what are called Jacks are used. These are simply pieces of vulcanite having brass plates on one side, the brass plates on two Jacks being connected to a flexible

cord. When the Jack is
forced between the spring
of the line wire and the
bottom plate, the latter is
connected to the telephone
by its upper contact, and at
the same time the Jack itself is in connection; and
when the other Jack is
forced between another
spring and its plate, the two
lines are in connection.

In the telephone ex-
changes in large towns, all
the indicators are on a large
frame; and operators, usually young ladies, are constantly employed noting the
indicators that drop, and
placing the jacks in position
to connect to the telephone
and to the subscriber re-
quired. In other parts of
the room are other oper-
ators, seated at tables, receiving the calls. A very
complete system of arrangement is necessary for des-
patch of business where the
calls are frequent. To an

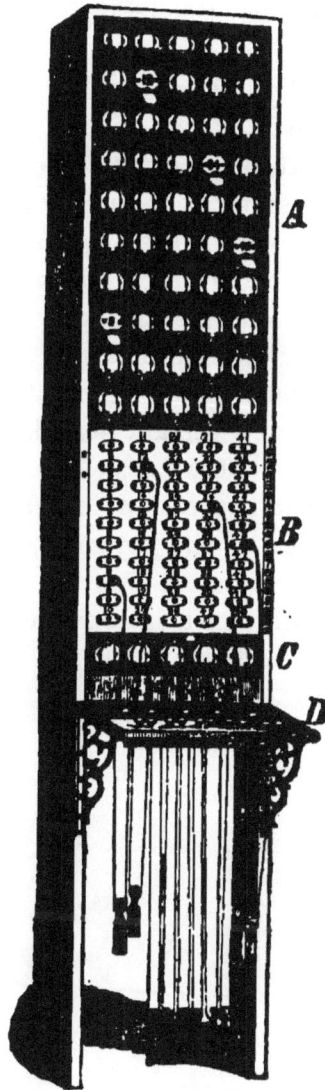

Fig. 127.—Telephone Switchboard.

outsider, the arrangements and the operations are very puzzling.

Fig. 127 shows one form of telephone exchange call-board.

Automatic call offices, which are now being used in Manchester and other towns, are worked something on the plan of the automatic cigarette boxes. The weight of a coin of a certain value makes contact and rings up the local centre, who in turn rings up a subscriber, when communication can go on. In some cases, the automatic arrangement cuts off communication at the end of a certain period, unless another spell is paid for.

Probably these automatic call boxes will only last until the call offices become sufficiently popular to admit of the payment of an attendant, and meanwhile they must necessarily be a source of some anxiety and trouble in maintenance, owing to their delicacy.

INDEX.

Butler & Tanner, The Selwood Printing Works, Frome, and London.

2, WHITE HART STREET,
PATERNOSTER SQUARE, E.C.

WHITTAKER & CO.'S
LIST OF
Classical, Educational, and Technical Works.

CONTENTS.

July, 1889.

Mr. Leland's Educational Publications.

Third Edition, Crown 8vo, Cloth, 6s.

PRACTICAL EDUCATION.

A WORK ON

PREPARING THE MEMORY, DEVELOPING QUICKNESS OF PERCEPTION, AND TRAINING THE CONSTRUC- TIVE FACULTIES.

By CHARLES G. LELAND.

Author of " The Minor Arts," " Twelve Manuals of Art Work," " The Album of Repoussé Work," Industrial Art in Education, or Circular No. 4, 1882," " Hints on Self-Education," etc.

MR. LELAND was the first person to introduce *Industrial Art* as a branch of education in the public schools of America. The Bureau of Education at Washington, observing the success of his work, employed him in 1862 to write a pamphlet showing how hand-work could be taken or taught in schools and families. It is usual to issue only 15,000 of these pamphlets, but so great was the demand for this that in two years after its issue more than 60,000 were given to applicants. This work will be found greatly enlarged in " Practical Education." Owing to it thousands of schools, classes, or clubs of industrial art were established in England, America and Austria. As at present a great demand exists for information as to organizing Technical Education, this forms the first part of the work. In it the author indicates that all the confusion and difference of opinion which at present prevails as to this subject, may very easily be obviated by simply beginning by teaching the youngest the easiest arts of which they are capable, and by thence gradually leading them on to more advanced work.

"The basis of Mr. Leland's theory," says a reviewer, "is that before learning, children should acquire the art of learning. It is not enough to fill the memory, memory must first be created. By training children to merely memorize, extraordinary power in this respect is to be attained in a few months. With this is associated exercises in quickness of perception, which are at first purely mechanical, and range from merely training the eye to mental arithmetic, and problems in all branches of education. Memory and quickness of perception blend in the development of the constructive faculties or hand-work. Attention or interest is the final factor in this system."

" *Mr. Leland's book will have a wide circulation. It deals with the whole subject in such a downright practical fashion, and is so much the result of long personal experience and observation, as to render it a veritable mine of valuable suggestions.*"—BRITISH ARCHITECT.

"*It has little of the dryness usually associated with such books ; and no teacher can read its thoughtful pages without imbibing many valuable ideas.*"—SCOTTISH EDUCATIONAL NEWS.

" *Strongly to be recommended.*"—CHEMICAL NEWS.

" *This valuable little work.*"—LIVERPOOL DAILY POST.

" *Many of Mr. Leland's suggestions might be carried out advantageously among the young folks in our large towns and villages.*"—NORTHERN WHIG.

𝔐inor 𝔄rts and 𝔍ndustries.

A SERIES OF ILLUSTRATED AND PRACTICAL MANUALS FOR

SCHOOL USE AND SELF-INSTRUCTION.

EDITED BY CHARLES G. LELAND.

This series of manuals on "The Minor Arts and Industries" is designed on the lines laid down in Mr. Leland's treatise on education. Each handbook will present the subject with which it deals in a thoroughly popular and practical manner; the lessons carry the student on his road step by step from the veriest elements to the point where the most advanced works fitly find their place in his course of study; in short, the greatest pains are taken to ensure a thorough mastery of the rudiments of each subject, and to so clearly state each lesson, illustrating it where necessary by plans and drawings, that even very young children may be interested in and trained to practical work. On similar grounds the self-taught student will find these manuals an invaluable aid to his studies.

Part 1 now ready, Paper cover, 1s. or in cloth, 1s. 6d.

DRAWING AND DESIGNING:

IN A SÉRIES OF LESSONS,

WITH NUMEROUS ILLUSTRATIONS,

By CHARLES G. LELAND, M.A., F.R.L.S.

Other volumes will follow at intervals, amongst the subjects of which may be named—

WOOD CARVING.	METAL WORK.
MODELLING.	CARPENTERING.
LEATHER WORK.	COMMERCE, ETC.

GREEK AND LATIN.

Bibliotheca Classica.

A Series of Greek and Latin Authors, with English Notes,
edited by eminent Scholars. 8vo.

ÆSCHYLUS. By F. A. Paley, M.A. 8*s.*

CICERO'S ORATIONS. By G. Long, M.A.
4 vols. 8*s.* each.

DEMOSTHENES. By R. Whiston, M.A. 2 vols.
8*s.* each.

EURIPIDES. By F. A. Paley, M.A. 3 vols. 8*s.* each.

HERODOTUS. By Rev. J. W. Blakesley, B.D. 2 vols. 12*s.*

HESIOD. By F. A. Paley, M.A. 5*s.*

HOMER. By F. A. Paley, M.A. Vol. I. 8*s.* Vol. II. 6*s.*

HORACE. By Rev. A. J. Macleane, M.A. 8*s.*

JUVENAL AND PERSIUS. By Rev. A. J.
Macleane, M.A. 6*s.*

LUCAN. The Pharsalia. By C. E. Haskins, M.A.,
and W. E. Heitland, M.A. 14*s.*

PLATO. By W. H. Thompson, D.D. 2 vols. 5*s.* each.

SOPHOCLES. Vol. I. By Rev. F. H. Blaydes, M.A. 8*s.*

——— Vol. II. Philoctetes—Electra—Ajax and Tra-
chiniæ. By F. A. Paley, M.A. 6*s.*

TACITUS: The Annals. By the Rev. P. Frost. 8*s.*

TERENCE. By E. St. J. Parry, M.A. 9*s.*

VIRGIL. By J. Conington, M.A. 3 vols. 10*s.* 6*d.* each.

** In some cases the volumes cannot be sold separately. The
few copies that remain are reserved for complete sets, which may be
obtained, at present, for 9*l.*

Grammar School Classics.

A Series of Greek and Latin Authors, with English Notes.
Fcap. 8vo.

CÆSAR: DE BELLO GALLICO.   By George
Long, M.A.   4*s.*

—— Books I.-III.   For Junior Classes.   By George
Long, M.A.   1*s. 6d.*

—— Books IV. and V. in 1 vol.   1*s. 6d.*

—— Books VI. and VII. in 1 vol.   1*s. 6d.*

CATULLUS, TIBULLUS, AND PROPER-
TIUS.   Selected Poems.   With Life.   By Rev. A. H.
Wratislaw.   2*s. 6d.*

CICERO: DE SENECTUTE, DE AMICITIA,
and SELECT EPISTLES.   By George Long, M.A.   3*s.*

CORNELIUS NEPOS.   By Rev. J. F. Macmichael.   2*s.*

HOMER: ILIAD.   Books I.-XII.   By F. A. Paley,
M.A.   4*s. 6d.*
Books I.-VI., 2*s. 6d.* ; Books VII.-XII., 2*s. 6d.*

HORACE.   With Life.   By A. J. Macleane, M.A.
3*s. 6d.*   In 2 Parts : Odes, 2*s.* ; Satires and Epistles, 2*s.*

JUVENAL: SIXTEEN SATIRES.   By H. Prior,
M.A.   3*s. 6d.*

MARTIAL: SELECT EPIGRAMS.   With Life.
By F. A. Paley, M.A.   4*s. 6d.*

OVID: The FASTI.   By F. A. Paley, M.A.   3*s. 6d.*

—— Books I. and II. in 1 vol.   1*s. 6d.*

—— Books III. and IV. in 1 vol.   1*s. 6d.*

—— Books V. and VI. in 1 vol.   1*s. 6d.*

SALLUST: CATILINA and JUGURTHA.   With
Life.   By G. Long, M.A., and J. G. Frazer.   3*s. 6d.* Catilina,
2*s.*   Jugurtha, 2*s.*

TACITUS: GERMANIA and AGRICOLA.   By
Rev. P. Frost.   2*s. 6d.*

VIRGIL : BUCOLICS, GEORGICS, and ÆNEID,
Books I.-IV. Abridged from Professor Conington's edition.
4*s.* 6*d.*

——— ÆNEID, Books V.-XII.   4*s.* 6*d.*
Also in 9 separate volumes, 1*s.* 6*d. each.*

XENOPHON: The ANABASIS. With Life. By
Rev. J. F. Macmichael. 3*s.* 6*d.*
Also in 4 separate volumes, 1*s.* 6*d. each.*

——— The CYROPÆDIA. By G. M. Gorham, M.A.
3*s.* 6*d.*

——— Books I. and II. in 1 vol.   1*s.* 6*d.*

——— Books V. and VI. in 1 vol.   1*s.* 6*d.*

——— MEMORABILIA. By Percival Frost, M.A.
3*s.*

**A GRAMMAR-SCHOOL ATLAS OF CLAS-
SICAL GEOGRAPHY,** containing Ten selected Maps.
Imperial 8vo. 3*s.*
Uniform with the Series.

THE NEW TESTAMENT, in Greek. With
English Notes, &c. By Rev. J. F. Macmichael. 4*s.* 6*d.*
Separate parts, St. Matthew, St. Mark, St. Luke, St. John,
Acts, 6*d.* each, sewed.

Lower Form Series.

With Notes and Vocabularies.

ECLOGÆ LATINÆ ; OR, FIRST LATIN
READING-BOOK, WITH ENGLISH NOTES AND A
DICTIONARY. By the late Rev. P. Frost, M.A. New
Edition. Fcap. 8vo. 1*s.* 6*d.*

**LATIN VOCABULARIES FOR REPETI-
TION.** By A. M. M. Stedman, M.A. 2nd Edition, revised.
Fcap. 8vo. 1*s.* 6*d.*

**EASY LATIN PASSAGES FOR UNSEEN
TRANSLATION.** By A. M. M. Stedman, M.A. Fcap.
8vo. 1*s.* 6*d.*

VIRGIL'S ÆNEID. Book I. Abridged from Conington's Edition by Rev. J. G. Sheppard, D.C.L. With Vocabulary by W. F. R. Shilleto. 1*s.* 6*d.* [*Now ready.*

CÆSAR : DE BELLO GALLICO. Book I. With Notes by George Long, M.A., and Vocabulary by W. F. R. Shilleto. 1*s.* 6*d.* [Book II. *in the press.*

TALES FOR LATIN PROSE COMPOSITION. With Notes and Vocabulary. By G. H. Wells, M.A. 2*s.*

MATERIALS FOR LATIN PROSE COMPOSITION. By the late Rev. P. Frost, M.A. New Edition. Fcap. 8vo. 2*s.* Key (for Tutors only), 4*s.*

A LATIN VERSE-BOOK. AN INTRODUCTORY WORK ON HEXAMETERS AND PENTAMETERS. By the late Rev. P. Frost, M.A. New Edition. Fcap. 8vo. 2*s.* Key (for Tutors only), 5*s.*

ANALECTA GRÆCA MINORA, with INTRODUCTORY SENTENCES, ENGLISH NOTES, AND A DICTIONARY. By the late Rev. P. Frost, M.A. New Edition. Fcap. 8vo. 2*s.*

GREEK TESTAMENT SELECTIONS. By A. M. M. Stedman, M.A. 2nd Edition, enlarged, with Notes and Vocabulary. Fcap. 8vo. 2*s.* 6*d.*

Cambridge Texts with Notes.

A Selection of the most usually read of the Greek and Latin Authors, Annotated for Schools. Fcap. 8vo, 1 s. 6d. each, with exceptions.

EURIPIDES. ALCESTIS — MEDEA — HIPPOLYTUS — HECUBA — BACCHÆ — ION (2*s.*) — ORESTES — PHOENISSÆ — TROADES – HERCULES FURENS — ANDROMACHE — IPHIGENIA IN TAURIS — SUPPLICES. By F. A. Paley, M.A., LL.D.

ÆSCHYLUS. PROMETHEUS VINCTUS — SEPTEM CONTRA THEBAS — AGAMEMNON — PERSÆ — EUMENIDES — CHŒPHORŒ. By F. A. Paley, M.A., LL.D.

SOPHOCLES. ŒDIPUS TYRANNUS—ŒDIPUS COLONEUS—ANTIGONE—ELECTRA—AJAX. By F. A. Paley, M.A., LL.D.

THUCYDIDES. BOOK IV. By. F. A. Paley, M.A., LL.D.

XENOPHON. HELLENICA. BOOK II. By Rev. L. D. Dowdall, M.A.

———— ANABASIS. Edited by Rev. J. F. Macmichael. *New edition*, revised by J. E. Melhuish, M.A. In 6 vols. Book I. (with Life, Introduction, Itinerary, &c.); Books II. and III. 2*s.* ; Book IV., Book V., Book VI., Book VII.

HOMER. ILIAD. BOOK I. By F. A. Paley, M.A., LL.D. 1*s.*

VIRGIL (abridged from Conington's edition). BUCO-LICS : GEORGICS, 2 parts : ÆNEID, 9 parts.

TERENCE. ANDRIA—HAUTON TIMORU-MENOS—PHORMIO—ADELPHOE. By Professor Wagner, Ph.D.

CICERO. DE SENECTUTE—DE AMICITIA—EPISTOLÆ SELECTÆ. By G. Long, M.A.

OVID. SELECTIONS. By A. J. Macleane, M.A.

Others in preparation.

Cambridge Greek and Latin Texts.

These Texts, which are clearly printed at the Cambridge University Press, on good paper, and bound in a handy form, have been reduced in price, and will now meet the requirements of masters who wish to use Text and Notes separately.

ÆSCHYLUS. By F. A. Paley, M.A. 2*s.*

CÆSAR: DE BELLO GALLICO. By G. Long, M.A. 1*s. 6d.*

CICERO: DE SENECTUTE et DE AMICITIA, et EPISTOLÆ SELECTÆ. By G. Long, M.A. 1*s. 6d.*

CICERONIS ORATIONES. Vol. I. (in Verrem.)
By G. Long, M.A. 2*s.* 6*d.*

EURIPIDES. By F. A. Paley, M.A. 3 vols., each 2*s.*

―――― Vol. I. Rhesus—Medea—Hippolytus—Alcestis—
Heraclidæ—Supplices—Troades—Index.

―――― Vol. II. Ion—Helena—Andromache—Electra—
Bacchæ—Hecuba—Index.

―――― Vol. III. Hercules Furens—Phœnissæ—Orestes—
Iphigenia in Tauris—Iphigenia in Aulide—Cyclops—Index.

HERODOTUS. By J. G. Blakesley, B.D. 2 vols.,
each 2*s.* 6*d.*

HOMERI ILIAS. I.-XII. By F. A. Paley, M.A.
1*s.* 6*d.*

HORATIUS. By A. J. Macleane, M.A. 1*s.* 6*d.*

JUVENAL ET PERSIUS. By A. J. Macleane,
M.A. 1*s.* 6*d.*

LUCRETIUS. By H. A. J. Munro, M.A. 2*s.*

SALLUSTI CRISPI CATILINA ET JU-
GURTHA. By G. Long, M.A. 1*s.* 6*d.*

SOPHOCLES. By F. A. Paley, M.A. 2*s.* 6*d.*

TERENTI COMŒDIÆ. By W. Wagner, Ph.D.
2*s.*

THUCYDIDES. By J. G. Donaldson, D.D. 2 vols.,
each 2*s.*

VERGILIUS. By J. Conington, M.A. 2*s.*

XENOPHONTIS EXPEDITIO CYRI. By
J. F. Macmichael, B.A. 1*s.* 6*d.*

NOVUM TESTAMENTUM GRÆCE. By
F. H. Scrivener, M.A. 4*s.* 6*d.* An edition with wide margin
for notes, half bound, 12*s.*

2 A

Annotated Editions.

CICERO'S MINOR WORKS. De Officiis, &c. &c. With English Notes, by W. C. Tylor, LL.D. 12mo. cloth, 3s. 6d.

VIRGIL'S ÆNEID. With English Notes, by C. Anthon, LL.D. Adapted for use in English Schools by the Rev. F. Metcalfe, M.A. *New Edition.* 12mo. 7s. 6d.

Greek Class Books.

BEATSON'S PROGRESSIVE EXERCISES ON THE COMPOSITION OF GREEK IAMBIC VERSE. 12mo. cloth, 3s.

DAWSON'S GREEK-ENGLISH LEXICON TO THE NEW TESTAMENT. *New Edition,* by Dr. Tylor. 8vo. cloth, 9s.

NOVUM TESTAMENTUM GRÆCE. Textus Stephanici, 1550. Accedunt variæ Lectiones editionum Bezæ, Elzeviri, Lachmanni, Tischendorfii, Tregellesii, curante F. H. Scrivener, M.A. 4s. 6d. An Edition with wide margin for MS. Notes, 4to. half-bound morocco, 12s.

—— Textûs Stephanici, A.D. 1550, Cum variis Lectionibus Editionum Bezæ, Elzeviri, Lachmanni, Tischendorfii, Tregellesii, Westcott-Hortii, Versionis Anglicanæ Emendatorum, Curante F. H. A. Scrivener, A.M., D.C.L., LL.D., Accedunt Parallela S. Scripturæ Loca. Small post 8vo. cloth, pp. xvi.-598, 7s. 6d.
EDITIO MAJOR *containing, in addition to the matter in the other Edition, the Capitula (majora et minora) and the Eusebian Canons, the various Readings of Westcott and Hort, and those adopted by the Revisers; also a revised and much-enlarged series of References.*

—— VALPY'S. For the use of Schools. 12mo. cloth, 5s.

—— Edited by Rev. Macmichael. *See Grammar School Classics.*

Students' Editions of the Gospels and the Acts.

Crown 8vo. cloth.

THE GOSPEL OF S. MATTHEW. The Greek Text, with Critical, Grammatical, and Explanatory Notes, &c., by the late Rev. W. Trollope, M.A., re-edited by the Rev. W. H. Rowlandson, M.A. 5s.

GOSPEL OF S. MARK. The Greek Text, with Critical, Grammatical, and Explanatory Notes, Prolegomena, &c., by Rev. W. H. Rowlandson, M.A. 4s. 6d.

GOSPEL OF S. LUKE. The Greek Text, with Critical, Grammatical, and Explanatory Notes, &c.,.by the late Rev. W. Trollope, M.A., revised and re-edited by the Rev. W. H. Rowlandson, M.A. 5s.

ACTS OF THE APOSTLES. The Greek Text, with Critical, Grammatical, and Explanatory Notes, and Examination Questions, by Rev. W. Trollope, M.A., re-edited and revised by the Rev. G. F. Browne, M.A. 5s.

Latin Class Books.

BEDFORD'S PROPRIA QUÆ MARIBUS; or, Short Rules for the Genders of Latin Nouns, and a Latin Prosody. 12mo. 1s.

BOSSUT'S LATIN WORD BOOK; or, First Step to the Latin Language. 18mo. 1s.

—— LATIN PHRASE BOOK. 18mo. 1s.

FLORILEGIUM POETICUM. A Selection of Elegiac Extracts from Ovid and Tibullus. *New edition,* greatly enlarged with English Notes. By the late Rev. P. Frost, M.A. Fcap. 8vo. 2s.

GRADUS AD PARNASSUM; sive novus sinonymorum, epithetorum, versuum, ac phrasium poeticarum, thesaurus. *New edition.* By G. Pyper. 12mo. cloth, 7s.

—— BY VALPY. Whittaker's Improved edition. Latin and English. *New edition.* Royal 12mo. 7s. 6d.

STODDART'S NEW DELECTUS; or, Easy Steps to Latin Construing. For the use of Pupils commencing the Language. Adapted to the best Latin Grammars, with a Dictionary attached. *New edition.* 12mo. 4s.

PENROSE'S (REV. JOHN) Easy Exercises in Latin Elegiac Verse. *New edition.* 12mo. cloth, 2s.

—— Key to ditto, for Tutors only, 3s. 6d.

Atlases.

LONG'S ATLAS OF CLASSICAL GEOGRAPHY. Containing Twenty-four Maps. Constructed by William Hughes, F.R.G.S., and Edited by George Long, M.A. *New edition*, with Coloured Outlines, and an Index of Places. Royal 8vo. 6s.

LONG'S GRAMMAR SCHOOL ATLAS OF CLASSICAL GEOGRAPHY. Containing Ten Maps, selected from the larger Atlas. Constructed by W. Hughes, F.R.G.S., and edited by George Long, M.A. *New edition*, with Coloured Outlines. Royal 8vo. 3s.

English Language and Miscellaneous.

ALLEN AND CORNWELL'S SCHOOL GRAMMAR. Cloth, 1s. 9d.

—— GRAMMAR FOR BEGINNERS. Cloth, 1s.

BELL'S MODERN READER AND SPEAKER. A Selection of Poetry and Prose, from the Writings of Eminent Authors. 12mo. 3s. 6d.

DUNCAN'S ENGLISH EXPOSITOR; or, Explanatory Spelling-book. Containing an Alphabetical Collection of all the most useful, proper, and elegant words in the English language, divided into Syllables, and properly accented. *New edition.* 12mo. 1s. 6d.

LATHAM'S (R. G.) DICTIONARY OF THE ENGLISH LANGUAGE. Abridged and condensed into one volume. 8vo. cloth, 14s.

MACKAY (C.) A DICTIONARY OF LOWLAND SCOTCH.

By Charles Mackay, LL.D. With an Introductory Chapter on the Poetry, Humour, and Literary History of the Scottish Language, and an Appendix of Scottish Proverbs. Large post 8vo. cloth, 7s. 6d. half bound, 8s. 6d.

—— SELECTED POEMS AND SONGS OF CHARLES MACKAY, LL.D. With a Commendatory and Critical Introduction by Eminent Writers. Wide foolscap 8vo. half cloth boards, 1s. 6d. Sewed, 1s.

WEBSTER'S DICTIONARY OF THE ENGLISH LANGUAGE.

Including Scientific, Technical, and Biblical Words and Terms. *New edition*, with Supplement of over 4,600 New Words and Meanings. 4to. cloth, 1l. 1s.; half-calf, 1l. 10s. With Appendices, £1 11s. 6d.; half-calf, 2l.

SHAKESPEARE'S PLAYS, with Text and Introduction in English and German.

Edited by C. Sachs, Prof. Ph. D. 8vo. cloth, each Play or Number, 10d.

Now Ready:

1. Julius Cæsar.
2. Romeo and Juliet.
3. King Henry VIII.
4. King Lear.
5. Othello.
6. Hamlet.
7. A Midsummer Night's Dream.
8. Macbeth.
9. King John.
10. King Richard II.
11. King Henry IV. I.
12. ,,   ,,   II.
13. King Henry V.
14. King Richard III.
15. Cymbeline.
16. Coriolanus.
17. Antony and Cleopatra.
18. Merchant of Venice.
19. Much Ado about Nothing.

Others to follow.

"This edition will be quite a godsend to grown-up students of either language, for the ordinary class reading books are too childish to arrest their attention. The parallel paging saves the labour of using a dictionary, and the series is so low in price as to place it within the reach of all."
Saturday Review.

SHAKESPEARE REPRINTS. 1. King Lear.

Parallel Texts of Quarto 1 and Folio 1. Edited by Dr. W. Vietor, of Marburg. Square 16mo. cloth, 3s. 6d.

The texts of the first quarto and folio, with collations from the later quartos and folios, are here printed in a compact and convenient volume, intended as a class-book in the University.

𝔐iscellaneous 𝔈ducational 𝔅ooks.

SEIDEL (ROBT.) INDUSTRIAL EDUCATION :
A Pedagogic and Social Necessity.   Crown 8vo. 4s.

WOODWARD (C.M.) THE MANUAL TRAIN-
ING SCHOOL, ITS AIMS, METHODS, AND RESULTS.
With Figured Drawings of Shop Exercises in Woods and Metals.
8vo. 10s.

BIBLIOGRAPHY OF EDUCATION.   Hints
toward a Select and Descriptive Bibliography of Education.
Arranged by Topics, and Indexed by Authors.   By G. Stanley
Hall, Professor, John Hopkins University, and John M.
Mansfield.   Post 8vo. cloth, pp. xvi.-309, 7s. 6d.

CHEPMELL'S (REV. DR.) SHORT COURSE
OF GRECIAN, ROMAN, AND ENGLISH HISTORY.
New edition.   12mo. 5s.   Questions on, 12mo. 1s.

COLTON (B. P.)   ELEMENTARY COURSE OF
PRACTICAL ZOOLOGY.   By B. P. Colton, A.M., Instructor
in Biology, Ottawa High School.   Crown 8vo. cloth, pp. xiv.-
182, 4s. 6d.

CORNWELL'S SCHOOL GEOGRAPHY.   3s. 6d.
With Thirty Maps on Steel, 5s. 6d.

—— GEOGRAPHY FOR BEGINNERS.   1s.
With Questions, 1s. 4d.

DURHAM UNIVERSITY CALENDAR, with
Almanack.   Cloth, 1s. 6d.   [*Published annually.*

JOYCE (P. W.) A HANDBOOK OF SCHOOL
MANAGEMENT AND METHODS OF TEACHING.
By P. W. Joyce, LL.D., &c.   11th edition, revised.   Cloth, 3s. 6d.

NATURE READERS.   Seaside and   Wayside.
No. 1.   By Julia McNair Wright.   Cloth, 1s. 6d.
   The first of a Series of Primary Readers for young children.
It treats of crabs, wasps, spiders, bees, and some univalve
molluscs.

PINNOCK'S HISTORY OF ENGLAND. From the Invasion of Julius Cæsar. With a Biographical and Historical Dictionary. Questions for Examination, Genealogical Tables, Progress of Literature and the Constitution, &c. Illusted. Continued by the Rev. W. H. Pinnock, LL.D. *New edition.* 12mo. 6s.

———— HISTORY OF GREECE. With an Introduction on the Natural and Political Geography of Greece, Dictionary of Difficult Terms, Questions for Examination, Genealogical Tables, &c. Illustrated. By Dr. W. C. Taylor. *New edition.* 12mo. 5s. 6d.

———— HISTORY OF ROME. With an Introduction, the Geography of the Roman Empire, Notices of the Roman Manners, and Illustrations, Questions for Examination, Chronological Index, &c. Illustrated. By Dr. W. C. Taylor. *New edition.* 12mo. 5s. 6d.

PINNOCK'S CATECHISMS OF THE ARTS, SCIENCES, AND LITERATURE. Whittaker's Improved Editions. Illustrated with Maps, Plates, and Woodcuts, carefully re-edited. 18mo. price 9d. each.

HISTORY.—Modern—Ancient—Universal—Bible and Gospel—Scripture— Chronology— England—Scotland— France—America—Rome—Greece—Jews.

GEOGRAPHY.—Ancient—Modern, Improved Edition—Modern, Original Edition—Sacred—England and Wales—Use of the Globes.

GRAMMAR.—English—French—German— Italian—Latin—Spanish—Greek : Part I. Accidence. Part II. Syntax and Prosody—Hebrew.

MATHEMATICS, &c.—Algebra (two Parts)—Arithmetic—Geometry—Navigation—Land Surveying.

RELIGION.—Religion—Natural Theology—Scripture History—Bible and Gospel History.

FINE ARTS, &c.—Architecture—Drawing—Perspective—Music—Singing.

LITERATURE.— Mythology— Rhetoric— Logic— British Biography—Classical Biography.

MISCELLANEOUS.—First Catechism—General Knowledge—Intellectual Philosophy—Agriculture—English Law—Heraldry—Medicine—Moral and Social Duties—Trade and Commerce.

SCHLEYER'S GRAMMAR, with Vocabularies of Volapuk (the Language of the World), for all Speakers of the English Language. Second (greatly Revised) Edition. By W. A. Seret, Certificated Teacher of the Universal Language. Crown 8vo. pp. 420, sewed, 5*s.* 6*d.* ; cloth, 6*s.* 6*d.*

SHUMWAY (E. S.) A DAY IN ANCIENT ROME. With numerous Illustrations. By Edgar S. Shumway, Professor, Rutger's College, New Brunswick. Small 4to. cloth, 5*s.*

WATTON'S ORIGINAL AIDS TO EDUCATION.

Hand-series of Tablets, in Stiff Covers, 3*d.* each.

Leading Events of General History.
Chief Events of Old Testament History.
Chief Events of New Testament History.
Prophecies and other Scripture Subjects.
Chief Events of Grecian History.
Chief Events of Roman History.
Chief Events of Eastern Empire.
Chief Events of German History.
Chief Events of English History.
Chief English Battles and Results.
Chief Events of Scottish History.
Chief Events of French History.
Chief Events of Prussian History.

Chief Events of Russian History.
Eminent Men of Modern Times.
Chief Events of Church History.
Natural System of Botany.
The Linnæan System of Botany.
Natural History—Zoology.
Natural Philosophy.
Principles of Grammatical Analysis, with Examples.
Guide to English Parsing, with Examples.
Abstract of Heathen Mythology.
Word Formation—Saxon, Latin, and Greek Prefixes, with Examples.
Chief Grecian and Roman Battles and Results.

LARGE TYPE SERIES OF TABLETS

(20 by 23 inches), embracing Historical, Geographical, and other Subjects, 4*d.* each, for suspension.

WATTON'S SKELETON EXERCISE BOOKS.

For History, Geography, Biography, Analysis, Parsing, and Chronology, with Script Headings and Specimen Page. Price regulated by the thickness of the books, 1*s.* and 2*s.* each. Also now ready, a filled Biographical Exercise Book, 2 Series, each 1*s.* Charts systematically arranged with date words, 60 pages, cloth, 1*s.* Selected Descriptive Poetry, 1*s.* Object Lessons, Nos. 1, 2, 3, and 4, 32 pp., in stiff covers, 2*d.* each.

School and University Analyses.

By the Rev. Dr. Pinnock.

AN ANALYSIS OF SCRIPTURE HIS-
TORY; Intended for Readers of Old Testament History, and
the University Examinations; with Maps, Copious Index, and
Examination Questions. 18mo. cloth, 3*s.* 6*d.*

AN ANALYSIS OF NEW TESTAMENT
HISTORY; Embracing the Criticism and Interpretation of
the original Text; with Questions for Examination. 18mo.
cloth, 4*s.*

AN ANALYSIS OF ECCLESIASTICAL
HISTORY; From the Birth of Christ, to the Council of
Nice, A.D. 325. With Examination Questions. 18mo. cloth,
3*s.* 6*d.*

ANALYSIS OF ENGLISH CHURCH HIS-
TORY; comprising the Reformation period, and subsequent
events; with Questions of Examination, especially intended for
the Universities and Divinity Students in general. 18mo. cloth,
4*s.* 6*d.*

A SHORT ANALYSIS OF OLD TESTA-
MENT HISTORY. With Questions for Schools. 18mo.
cloth, 1*s.* 6*d.*

A SHORT ANALYSIS OF NEW TESTA-
MENT HISTORY. With Questions for Schools. 18mo.
cloth, 1*s.* 6*d.*

Arithmetic and Euclid.

PINNOCK'S ARITHMETICAL TABLES OF
MONEY, WEIGHTS, AND MEASURES. With Questions
for Examination, and Explanatory Notes, &c. 18mo. 3*d.*

—— FIRST CIPHERING BOOK. Containing
Easy Exercises in the First Rules of Arithmetic. 4to. sewed, 1*s.*

RYAN'S CIVIL SERVICE ARITHMETICAL
EXAMINATION PAPERS. By L. J. Ryan. Cloth, 2*s.*

—— Key to Ditto. 1*s.* 6*d.*

SONNENSCHEIN AND NESBITT'S
ARITHMETIC. The Science and Art of Arithmetic for the use of Schools. Post 8vo. 5s. 6d. Or separately, Part I.—Integral. 2s. 6d. Parts II. and III.—Fractional and Approximate Calculations. 3s. 6d. Answers to the Exercises. 1s. 6d. Exercises separately. Part I. 1s. Parts II. and III. 1s. 3d.

—— A B C OF ARITHMETIC. Teacher's Book, Nos. 1 and 2, each 1s. Exercise Book, Nos. 1 and 2, each 4d.

WALKINGAME'S TUTOR ASSISTANT (FRASER'S). Being a Compendium of Arithmetic and a Complete Question Book. 12mo. 2s. Key, 3s.

EUCLID, THE FIRST BOOK OF. With an Introduction and Collection of Problems for the use of Schools. By J. M. Wilson, M.A. *2nd edition.* 4to. 2s.

EUCLID, THE FIRST SIX BOOKS, together with the ELEVENTH and TWELFTH. From the Text of Dr. Simson. *New edition*, revised and corrected by S. Maynard. 18mo. 4s.

MODERN LANGUAGES.

𝔉rench.

BARRÈRE (A.) PROFESSOR, R.M.A. Woolwich.

—— RÉCITS MILITAIRES. Selections from modern French authors, with short biographical introductions in French, and English notes for the use of army students and others. Crown 8vo. 3s.

—— PRÉCIS OF COMPARATIVE FRENCH GRAMMAR AND IDIOMS, and Guide to Examinations. Cloth. *Second edition, revised,* 3s. 6d.

—— JUNIOR GRADUATED FRENCH COURSE affording materials for Translation, Grammar, and Conversation. Being an introduction to the Graduated French Course. Cloth, 1s. 6d.

—— ELEMENTS OF FRENCH GRAMMAR AND FIRST STEPS IN IDIOMS. With numerous Exercises and a Vocabulary, being an Introduction to the Précis of Comparative French Grammar. Crown 8vo. cloth, 2s.

BELLENGER'S MODERN FRENCH CONVER-
SATION. Containing Elementary Phrases and New Easy
Dialogues, in French and English, on the most familiar sub-
jects. 12mo. 2*s.* 6*d.*

BOSSUT'S FRENCH WORD BOOK. 18mo. 1*s.*

——— FRENCH PHRASE BOOK. 18mo. 1*s.*

BOWER. PUBLIC EXAMINATION FRENCH
READER. With a Vocabulary to every extract, suitable for
all Students who are preparing for a French Examination. By
A. M. Bower, F.R.G.S., late Master in University College
School, &c. Cloth, 3*s.* 6*d.*

"The book is a very practical and useful one, and it must prove very handy
for students who are preparing for a French examination, the persons for whose
special aid it has been specially provided. It would also serve admirably for use
in schools as a class book."—*Schoolmaster.*

DELILLE'S FRENCH GRAMMAR. In Two
Parts. I.—Accidence. II.—Syntax, written in French, with
Exercises conducive to the speaking of the French Language,
&c. 12mo. 5*s.* 6*d.* Key, 3*s.*

——— EASY FRENCH POETRY FOR BEGIN-
NERS ; or, Short Selections in Verse on a Graduated Plan for
the Memory. With English Notes. 12mo. 2*s.*

DELILLE'S MODÈLES DE POÉSIE FRANÇAIS.
With Treatise on French Versification. *New edition.* 12mo.
6*s.*

——— RÉPERTOIRE DES PROSATEURS FRAN-
CAIS. With Biographical Sketches, &c. *New edition.*
12mo. 6*s.* 6*d.*

——— MANUEL ÉTYMOLOGIQUE ; or, an Inter-
pretative Index of the most recurrent Words in the French
Language. 12mo. 2*s.* 6*d.*

——— BEGINNER'S OWN FRENCH BOOK.
Being a Practical and Easy Method of Learning the Elements
of the French Language. 12mo. cloth, 2*s.* Key, 2*s.*

DES CARRIÈRES' FRENCH IDIOMATICAL
PHRASES AND FAMILIAR DIALOGUES. Square,
3*s.* 6*d.*

DES CARRIÈRES' HISTOIRE DE FRANCE, DEPUIS L'ETABLISSEMENT DE LA MONARCHIE. Continuée jusqu'au rétablissement de l'Empire sous Napoleon III., par C. J. Delille. 12mo. 7s.

DUVERGER'S COMPARISON BETWEEN THE IDIOMS, GENIUS, AND PHRASEOLOGY OF THE FRENCH AND ENGLISH LANGUAGES. *New edition.* 12mo. 4s. 6d.

GASC (F. E. A.) AN IMPROVED MODERN POCKET DICTIONARY OF THE FRENCH AND ENGLISH LANGUAGES. *New edition.* 16mo. cloth, 2s. 6d. Also in 2 vols. in neat leatherette, 5s.

—— MODERN FRENCH-ENGLISH AND ENGLISH-FRENCH DICTIONARY. *New edition, revised.* In 1 vol. 8vo. 10s. 6d.

HAMEL'S NEW UNIVERSAL FRENCH GRAMMAR. *New Edition.* 12mo. 4s.

—— GRAMMATICAL EXERCISES UPON THE FRENCH LANGUAGE. *New edition.* 12mo. 4s. Key, 3s.

—— FRENCH GRAMMAR AND EXERCISES. *New edition.* 12mo. 5s. 6d. Key, 4s.

LEVIZAC'S DICTIONARY OF THE FRENCH AND ENGLISH LANGUAGES. *New edition*, by N. Lambert. 12mo. 6s. 6d.

NUGENT'S POCKET DICTIONARY OF THE FRENCH AND ENGLISH LANGUAGES. *New edition*, revised by J. C. J. Tarver. *Pearl edition*, 4s. 6d.

OLLENDORFS (Dr. H. G.) NEW METHOD OF LEARNING TO READ, WRITE, AND SPEAK A LANGUAGE IN SIX MONTHS. Adapted to the French. *New edition.* 12mo. 6s. 6d. Key, 8vo. 7s.

PRACTICAL COMMERCIAL CORRESPON-DENCE. *See Miscellaneous.*

𝔚𝔥𝔦𝔱𝔱𝔞𝔨𝔢𝔯'𝔰 𝔉𝔯𝔢𝔫𝔠𝔥 𝔖𝔢𝔯𝔦𝔢𝔰.

For the use of Schools and Private Students. Edited by
A. Barrère, Prof. R.M.A. Woolwich, &c., and others. Each
number with a literary Introduction and Arguments in English,
foot-notes explaining the more difficult passages, and translations
of the idiomatic expressions into the corresponding English idioms.

Fcap. 8vo, each number, sewed, 6d. ; cloth, 9d.

Now Ready :—

1. SCRIBE. LE VERRE D'EAU. Barrère.
2. MOLIERE. LE BOURGEOIS GENTILHOMME. Gasc.
3. MOLIERE. L'AVARE. Gasc.
4. SOUVESTRE. SOUS LA TONNELLE. Desages.
5. MOLIERE. LE MISANTHROPE. Gasc.
6. GALLAND. ALI BABA. Clare.
7. CORNEILLE. LE CID. Gasc.
8, 9. LAMARTINE. JEANNE D'ARC. Barrère.
10, 11. PIRON. LA METROMANIE. Delbos.

Others to follow.

𝔚𝔥𝔦𝔱𝔱𝔞𝔨𝔢𝔯'𝔰 𝔉𝔯𝔢𝔫𝔠𝔥 𝔠𝔩𝔞𝔰𝔰𝔦𝔠𝔰, 𝔴𝔦𝔱𝔥 𝔈𝔫𝔤𝔩𝔦𝔰𝔥 𝔑𝔬𝔱𝔢𝔰.

Fcap. 8vo. cloth.

AVENTURES DE TELEMAQUE. Par Féné-
lon. *New edition.* Edited and revised by C J. Delille. 2s. 6d.

HISTOIRE DE CHARLES XII. Par Voltaire.
New edition. Edited and revised by L. Direy. 1s. 6d.

PICCIOLA. Par X. B. Saintine. *New edition.* Edited
and revised by Dr. Dubuc. 1s. 6d.

SELECT FABLES OF LA FONTAINE. *New
edition.* Edited by F. Gasc, M.A. 1s. 6d.

𝔚hittaker's Series of 𝔐odern French Authors.

WITH INTRODUCTION AND NOTES.

For Beginners.

LA BELLE NIVERNAISE. Histoire d'un vieux bateau et de son équipage. By Alphonse Daudet. With 6 illustrations. Edited by James Boïelle, Senior French Master at Dulwich College. 2*s*. 6*d*.   [*Ready.*

For Advanced Students.

BUG JARGAL. By Victor Hugo. Edited by James Boïelle, Senior French Master at Dulwich College. 3*s*.
Others to follow.   [*Ready.*

German.

FLÜGEL'S COMPLETE DICTIONARY OF THE GERMAN AND ENGLISH LANGUAGES. Comprising the German and English, and English and German. Adapted to the English Student, with great Additions and Improvements. By C. A. Feiling, A. Heimann, and J. Oxenford. *New edition.* 2 vols. 8vo. 1*l*. 1*s*.

———— ABRIDGED GERMAN AND ENGLISH, AND ENGLISH AND GERMAN DICTIONARY. Carefully compiled from the larger Dictionary. By C. A. Feiling and J. Oxenford. *New edition.* Royal 18mo. 6*s*.

GRENFELL'S ELEMENTARY GERMAN EXERCISES. Part I. Adapted to the Rugby School German Accidence. 12mo. 1*s*. 6*d*.

OLLENDORFF'S (Dr. H. S.). NEW METHOD OF LEARNING TO READ, WRITE, AND SPEAK A LANGUAGE IN SIX MONTHS. Adapted to the German *New edition.* Crown 8vo. 7*s*. Key, 8vo. 7*s*.

SHELDON (E.S.) A SHORT GERMAN GRAMMAR FOR HIGH SCHOOLS AND COLLEGES. Crown 8vo. 3*s*.

WHITTAKER'S COURSE OF MODERN GERMAN.
By F. Lange, Ph.D., Professor, R.M.A. Woolwich, Examiner in German to the College of Preceptors, London; Examiner in German at the Victoria University, Manchester, and J. F. Davis, M.A., D.Lit. Extra fcap. 8vo. cloth.

A CONCISE GERMAN GRAMMAR. With especial reference to Phonology, Comparative Philology, English and German Correspondences, and Idioms. By Frz. Lange, Ph.D., Professor at the Royal Military Academy, Woolwich. In three Parts. Part I., Elementary, 2s. Part II., Intermediate, 2s. Part III. in the press.

ELEMENTARY GERMAN READER. A Graduated Collection of Readings in Prose and Poetry. With English Notes and a Vocabulary. By F. Lange, Ph.D. 1s. 6d.

ADVANCED GERMAN READER. A Graduated Collection of Readings in Prose and Poetry. With English Notes and a Vocabulary. By F. Lange, Ph.D. and J. F. Davis, M.A., D.Lit. *[Nearly ready.*

PROGRESSIVE GERMAN EXAMINATION COURSE. In Three Parts. By F. Lange, Ph.D., Prof. R.M.A., Woolwich, Examiner in German to the College of Preceptors.

Comprising the Elements of German Grammar, an Historical Sketch of the Teutonic Languages, English and German Correspondences, Materials for Translation, Dictation, Extempore, Conversation and complete Vocabularies.

1. ELEMENTARY COURSE. Cloth, 2s.
2. INTERMEDIATE COURSE. Cloth, 2s.
3. ADVANCED COURSE. Second revised edition. Cloth, 1s. 6d.

"We cordially commend it as a useful help to examiners, who will find it well adapted to their needs."—*Practical Teacher.*

In Preparation.

GERMAN CONVERSATIONAL DICTIONARY.
A Guide to Modern German Conversation. For the Every-day Purposes of Travellers and Students. By G. May and J. F. Davis, M.A., D.Lit.

German Classics, with English Notes.

Fcap. 8vo. cloth.

GERMAN BALLADS. From Uhland, Goethe, and Schiller. With Introductions to each Poem, copious Explanatory Notes, and Biographical Notices. By C. Bielefeld. 1*s.* 6*d.*

GOETHE'S HERMANN AND DOROTHEA. With Short Introduction, Argument, and Notes Critical and Explanatory. By Ernest Bell and E. Wölfel. 1*s.* 6*d.*

SCHILLER'S MAID OF ORLEANS. With Introduction and Notes. By Dr. Wagner. 1*s.* 6*d.*

—— **MARIA STUART.** With Introduction and Notes. By V. Kastner, M.A. 1*s.* 6*d.*

—— **WALLENSTEIN.** Complete Text. *New edition.* With Notes, Arguments, and an Historical and Critical Introduction. By C. A. Buchheim, Professor, Ph.D., 5*s.* Or separately—Part I.—THE LAGER AND DIE PICCOLOMINI. 2*s.* 6*d.* Part II.—WALLENSTEIN'S TOD. 2*s.* 6*d.*

Whittaker's Series of Modern German Authors.

With Introduction and Notes. Edited by F. Lange, Ph.D., Professor, Royal Military Academy, Woolwich.

The attention of the heads of Colleges and Schools is respectfully directed to this new Series of " MODERN GERMAN AUTHORS " which is intended to supply the much-felt want of suitable Reading Books for English Students of German who have passed through the preliminary stages of fables and anecdotes.

To those who wish to extend their linguistic and grammatical knowledge, these volumes will afford, in one respect, a great advantage over those of an earlier period, presenting, as they do, the compositions of the best living, or only recently deceased authors. The Notes, besides etymological and other explanations, will contain many useful idiomatic expressions suggested by the text, and worth committing to memory.

FIRST SERIES.

FOR BEGINNERS. Edited, with a Grammatical Introduction, Notes, and a Vocabulary, by F. Lange, Ph. D., Professor, R. M. A. Woolwich, Examiner in German to the College of Preceptors, and H. Hager, Ph. D., Examiner in German to the London University.

HEY'S FABELN FÜR KINDER. Illustrated by O. Speckter. Edited, with an Introduction, Grammatical Summary, Words, and a complete Vocabulary. By F. Lange, Ph. D., Professor. 1s. 6d.

THE SAME, with a Phonetic Introduction, Phonetic Transcription of the Text. By F. Lange, Professor, Ph. D. 2s.

SECOND SERIES.

FOR INTERMEDIATE STUDENTS. Edited, with a Biographical Introduction, Notes, and a complete vocabulary, by F. Lange, Ph. D., Professor, and H. Hager, Ph. D.

DOKTOR WESPE. Lustspiel in fünf Aufzügen von JULIUS RODERICH BENEDIX. Edited by F. Lange, Ph. D., Professor. 2s. 6d.

SCHILLER'S JUGENDJAHRE. Erzählung von FRZ. HOFF-MANN. Edited by H. Hager, Ph. D., Professor. [*In the press.*

THIRD SERIES.

FOR ADVANCED STUDENTS. Edited, with a Literary Introduction and Notes, by F. Lange, Ph. D., Professor, R. M. A. Woolwich, in co-operation with F. Storr, B.A. ; A. A. Macdonell, M.A. ; H. Hager, Ph. D. ; C. Neuhaus, Ph. D. and others.

MEISTER MARTIN, der Küfner. Erzählung von E. T. A. Hoffman. Edited by F. Lange, Ph. D., Professor, Royal Military Academy, Woolwich. 1s. 6d.

HANS LANGE. Schauspiel von Paul Heyse. Edited by A. A. Macdonell, M.A., Ph.D., Taylorian Teacher, University, Oxford. 2s.

AUF WACHE. Novelle von Berthold Auerbach. DER GEFRO-RENE KUSS. Novelle von Otto Roquette. Edited by A. A. Macdonell, M.A. 2s.

DER BIBLIOTHEKAR. Lustspiel von G. von Moser. Edited by F. Lange, Ph.D. Second revised Edition. 2s.

EINE FRAGE. Idyll von George Ebers. Edited by F. Storr, B.A., Chief Master of Modern Subjects in Merchant Taylor's School. 2s.

DIE JOURNALISTEN. Lustspiel von Gustav Freytag. Edited by Professor F. Lange, Ph.D. Second revised Edition. 2s. 6d.

ZOPF UND SCHWERT. Lustspiel von Karl Gutzkow. Edited by Professor F. Lange, Ph.D. 2s. 6d.

GERMAN EPIC TALES IN PROSE. I. Die Nibelungen, von A. F. C. Vilmar.—II. Walther und Hildegund, von Albert Richter. Edited by Karl Neuhaus, Ph.D., the International College, Isleworth. 2s. 6d.

Italian.

BARETTI'S DICTIONARY OF THE ENGLISH AND ITALIAN LANGUAGES. To which is prefixed an Italian and English Grammar. *New Edition,* entirely re-written. By G. Comelati and J. Davenport. 2 vols. 8vo. 1l. 1s.

GRAGLIA'S NEW POCKET DICTIONARY OF THE ITALIAN AND ENGLISH LANGUAGES. With considerable Additions, and a Compendious Elementary Italian Grammar. 18mo. 4s. 6d.

OLLENDORFF'S (DR. H. G.) NEW METHOD OF LEARNING TO READ, WRITE, AND SPEAK A LANGUAGE IN SIX MONTHS. Adapted to the Italian. *New Edition.* Crown 8vo. 7s. Key, 8vo. 7s.

SOAVE'S NOVELLE MORALI. *New Edition.* 12mo. 4s.

VENERONI'S COMPLETE ITALIAN GRAM-
MAR. By P. Rosteri. 12mo. 6s.

VERGANI AND PIRANESI'S ITALIAN AND
ENGLISH GRAMMAR. With Exercises, &c. By J.
Guichet. *New edition,* by Signor A. Tommasi. 12mo. 5s.
Key, 3s.

Russian.

DOLBESHOFF (E.) A DICTIONARY OF THE
RUSSIAN AND ENGLISH LANGUAGES. In two
volumes. Vol. I. Russian-English. Vol. II. English-Russian.
Compiled by E. Dolbeshoff in co-operation with C. E. Turner,
Professor of English Language and Literature at the University,
St. Petersburg. [*Preparing.*

Spanish.

NEUMAN AND BARETTI'S SPANISH AND
ENGLISH, AND ENGLISH AND SPANISH DIC-
TIONARY. Revised and enlarged by M. Seoane, M.D.
2 vols. 8vo. 1l. 8s.

—— POCKET DICTIONARY. Spanish and Eng-
lish, and English and Spanish. Compiled from the larger
work. 18mo. 5s.

OLLENDORFF'S (DR. H. G.) NEW METHOD
OF LEARNING TO READ, WRITE, AND SPEAK A
LANGUAGE IN SIX MONTHS. Adapted to the Spanish.
New edition. 8vo. 12s. Key, 8vo. 7s.

PONCE DE LEON'S ENGLISH - SPANISH
TECHNOLOGICAL DICTIONARY. 8vo. 1l. 16s. *See*
page 31.

Practical Mercantile Correspondence.

A Collection of Commercial Letters and Forms, with Notes, Explanatory and Grammatical, and a Vocabulary of Commercial Terms, edited by L. Simon, Chr. Vogel, Ph.D., H. P. Skelton, W. C. Wrankmore, Leland Mason, and others. Intended as Class Books for Schools and for Self-Instruction.

Now Ready, crown 8vo, cloth :
ENGLISH, with German Notes, 3*s.*
GERMAN, with English Notes, 3*s.*
ENGLISH, with French Notes, 4*s.* 6*d.*
FRENCH, with English Notes, 4*s.* 6*d.*

This new Collection of Model Letters and Epistolary Forms embraces the whole sphere of Commercial Transactions. Each example is provided with such remarks and explanations, that any one with a fair grammatical knowledge of the particular language will find it an easy matter to prepare a well-expressed letter.

The Specialist's Series.

A New Series of Handbooks for Students and Practical Engineers. Crown 8vo. With many Illustrations.

GAS ENGINES. Their Theory and Management. By William Macgregor. With 7 Plates. Crown 8vo. pp. 245, 8*s.* 6*d.*

BALLOONING: A Concise Sketch of its History and Principles. From the best sources, Continental and English. By G. May. With Illustrations. Crown 8vo. pp. vi.-97, 2*s.* 6*d.*

ELECTRIC TRANSMISSION OF ENERGY, and its Transformation, Subdivision, and Distribution. A Practical Handbook by Gisbert Kapp, C.E., Associate Member of the Institution of Civil Engineers, &c. With 119 Illustrations. Crown 8vo. pp. xi.-331. 7*s.* 6*d.*

ARC AND GLOW LAMPS. A Practical Handbook on Electric Lighting. By Julius Maier, Ph.D., Assoc. Soc. Tel. Eng., &c. With 78 Illustrations. Crown 8vo. pp. viii.-376. 7*s.* 6*d.*

ON THE CONVERSION OF HEAT INTO
WORK. A Practical Handbook on Heat-Engines. By William Anderson, M. Inst. C.E. With 64 Illustrations. Pp. viii.-254. Cr. 8vo. 6s.

SEWAGE TREATMENT, PURIFICATION
AND UTILIZATION ; A Practical Manual for the Use of Corporations, Local Boards, Officers of Health, Inspectors of Nuisances, Chemists, Manufacturers, Riparian Owners, Engineers and Ratepayers. By J. W. Slater, F.E.S., Editor of "Journal of Science." Crown 8vo. cloth, price 6s.

THE TELEPHONE. By W. H. Preece, F.R.S.,
and J. Maier, Ph.D. With numerous illustrations. Cr. 8vo. 12s. 6d.

MANURES, OR THE PHILOSOPHY OF
MANURING. By Dr. A. B. Griffiths, F.R.S.Ed., F.C.S., Principal and Lecturer on Chemistry in the School of Science, Lincoln, &c., &c. Cr. 8vo. 7s. 6d.

HYDRAULIC MOTORS: TURBINES AND
PRESSURE MOTORS. By George R. Bodmer, Assoc. M.Inst.C.E. 14s.

ALTERNATING CURRENTS OF ELEC-
TRICITY. By Thomas H. Blakesley, M.A., M.Inst.C.E. 4s. 6d.

In preparation.

GALVANIC BATTERIES. By Professor George Forbes, M.A.

INDUCTION COILS. By Professor A. J. Fleming, M.A., D.Sc.

THE DYNAMO. By Guy C. Fricker.

Others to follow.

NIPHER (F. E.) THEORY OF MAGNETIC
MEASUREMENTS, WITH AN APPENDIX ON THE METHOD OF LEAST SQUARES. One volume. Crown 8vo. cloth, 5s.

PLANTÉ (G.) THE STORAGE OF ELECTRI-
CAL ENERGY, and Researches in the Effects created by Currents combining Quantity with High Tension. Translated from the French by Paul Bedford Elwell. With Portrait, and 89 Illustrations. 8vo. pp. vii.-268, cloth, 12s.

Small crown 8vo. cloth.   With many Illustrations.

𝔚𝔥𝔦𝔱𝔱𝔞𝔨𝔢𝔯'𝔰 𝔏𝔦𝔟𝔯𝔞𝔯𝔶 𝔬𝔣 𝔄𝔯𝔱𝔰, 𝔖𝔠𝔦𝔢𝔫𝔠𝔢𝔰, 𝔐𝔞𝔫𝔲𝔣𝔞𝔠𝔱𝔲𝔯𝔢𝔰 𝔞𝔫𝔡 𝔍𝔫𝔡𝔲𝔰𝔱𝔯𝔦𝔢𝔰.

MANAGEMENT OF ACCUMULATORS AND PRIVATE ELECTRIC LIGHT INSTALLATIONS. A Practical Handbook by Sir David Salomons, Bart., M.A. 4th Edition, Revised and Enlarged, with 32 Illustrations. Cloth, 3s.

"To say that this book is the best of its kind would be a poor compliment, as it is practically the only work on accumulators that has been written."—*Electrical Review.*

ELECTRICAL INSTRUMENT-MAKING FOR AMATEURS. A Practical Handbook. By S. R. Bottone, Author of "The Dynamo," &c. With 60 Illustrations. Third edition. Cloth, 3s.

ELECTRIC BELLS AND ALL ABOUT THEM. A Practical Book for Practical Men. By S. R. Bottone. With more than 100 illustrations. Cloth, 3s.

PRACTICAL IRON FOUNDING. By the Author of "Pattern Making," &c., &c. Illustrated with over one hundred engravings. Cloth, 4s.

THE PROTECTION OF BUILDINGS FROM LIGHTNING. A Treatise on the Theory of Lightning Conductors from a Modern Point of View. Being the substance of two lectures delivered before the Society of Arts in March, 1888. By Oliver J. Lodge, LL.D., D.Sc., F.R.S., Professor of Physics in University College, Liverpool.

Published with various amplifications and additions, with the approval of the Society of Arts.   [*In preparation.*

ELECTRICAL INFLUENCE MACHINES: Containing a full account of their historical development, their modern Forms, and their Practical Construction. By J. Gray. B.Sc.   [*In the press.*

ELECTRICAL ENGINEERING IN OUR HOMES AND WORKSHOPS. A Practical Handbook. By Sydney F. Walker, M.Inst.C.E., M.I.E.E. [*Immediately.*
Others in preparation.

Technological Dictionaries.

ENGLISH AND GERMAN.

WERSHOVEN (F. J.) TECHNOLOGICAL DICTIONARY OF THE PHYSICAL, MECHANICAL, AND CHEMICAL SCIENCES. English and German. 2 vols. cloth, 5s.

ENGLISH—SPANISH.

PONCE DE LEON. TECHNOLOGICAL DICTIONARY. English-Spanish and Spanish-English. Containing Terms employed in the Applied Sciences, Industrial Arts, Mechanics, Fine Arts, Metallurgy, Machinery, Commerce, Ship-building and Navigation, Civil and Military Engineering, Agriculture, Railway Construction, Electro-technics, &c.

Vol. I.—English-Spanish. 8vo. bound, £1 16s.

Vol. II.—Spanish-English. [*In preparation.*

Post 8vo. 814 pp. 10s. 6d.

HOBLYN'S DICTIONARY OF TERMS USED IN MEDICINE AND COLLATERAL SCIENCES. 11*th edition.* Revised throughout, with numerous Additions. By John A. P. Price, B.A., M.D. Oxon., Assistant-Surgeon to the Royal Berkshire Hospital.

This new edition has undergone complete revision and emendation. Many terms, fallen more or less into disuse, have been omitted; and a considerable amount of fresh matter has been introduced, in order to meet the requirements of the present day.

1 vol. demy 4to. with 25 Double and 40 Single Plates, £2 10s.

TECHNICAL SCHOOL AND COLLEGE BUILDING.

*Being a Treatise on the Design and Construction of
Applied Science and Art Buildings, and
their suitable Fittings and
Sanitation.*

WITH A CHAPTER ON TECHNICAL EDUCATION.

By EDWARD COOKWORTHY ROBINS, F.S.A.

OUTLINE OF CONTENTS.—Introduction—English and Foreign Technical Education—Analysis of the Second Report of the Royal Commissioners on Technical Education—Buildings for Applied Science and Art Instruction, with examples of Foreign and English Buildings—Analysis of the Fittings necessary for these Buildings— British and Foreign Examples of the Details of the Fittings—Heating and Ventilation generally—Heating and Ventilation necessary for Applied Science and Instruction Buildings—The Planning of Buildings for Middle Class Education—Sanitary Science—Appendix.

Full prospectus post free on application.

"It will prove an indispensable work of reference to architects, builders, and managers of technical schools."—*Spectator.*
"A most valuable contribution to architectural literature."—*British Architect.*

THE CHEMICAL ANALYSIS OF IRON.

A Complete Account of all the Best Known Methods for the Analysis of Iron, Steel, Ores, &c.

By A. A. BLAIR, Chief Chemist, U.S. Geological Survey, &c.
Royal 8vo.   14s.

Just published.

THE

WORKING AND MANAGEMENT OF AN ENGLISH RAILWAY.

By GEORGE FINDLAY,

General Manager of the London and North-Western Railway.

WITH NUMEROUS ILLUSTRATIONS.

Crown 8vo.   7s. 6d.

CHISWICK PRESS:—C. WHITTINGHAM AND CO., TOOKS COURT,
CHANCERY LANE.

www.ingramcontent.com/pod-product-compliance
Lightning Source LLC
Chambersburg PA
CBHW030914270326
41929CB00008B/689